D1268479

Fire in the Sea

The Santorini Volcano: Natural History and the Legend of Atlantis

When the Greek Island of Santorini, classically known as Thera, erupted dramatically in the seventeenth century BC, it produced one of the largest explosions ever witnessed by humankind. The event covered Bronze Age settlements on the island with volcanic ash, and altered the course of civilization in the region, possibly giving rise to the legend of Atlantis. The ejection of an immense volume of dust into the atmosphere also altered global temperatures for several years.

Fire in the Sea blends the thrill of scientific discovery with a popular presentation of the geology, archaeology, history, peoples, and environmental setting of Santorini. Walter Friedrich has worked for many years on the Minoan eruption of the island. He discusses the latest scientific results and combines natural history and archaeology to provide a comprehensive overview of the volcanic island and its past. He discusses Plato's legend of Atlantis, and other myths that are often associated with Santorini. The book is a fascinating case study of an environmental disaster which had a strong impact on ancient history and created a 'Bronze Age Pompeii'.

Excellent color photographs and illustrations along with easily understandable scientific and historic details make this book highly appealing to a wide audience. It will also be useful as a supplementary text for introductory courses in earth and atmospheric science, geology, volcanology, paleoclimatology, as well as ancient history and archaeology.

Caldera wall below the town of Fira on Thera

Fire in the Sea

The Santorini Volcano: Natural History and the Legend of Atlantis

WALTER L FRIEDRICH

TRANSLATED BY ALEXANDER R. McBIRNEY

CAMBRIDGE
UNIVERSITY PRESS

PUBLISHED BY THE PRESS SYNDICATE OF THE UNIVERSITY OF CAMBRIDGE
The Pitt Building, Trumpington Street, Cambridge, United Kingdom

CAMBRIDGE UNIVERSITY PRESS
The Edinburgh Building, Cambridge CB2 2RU, UK http://www.cup.cam.ac.uk
40 West 20th Street, New York, NY 10011-4211, USA http://www.cup.org
10 Stamford Road, Oakleigh, Melbourne 3166, Australia
Ruiz de Alarcón 13, 28014 Madrid, Spain

First published in German as *Feuer im Meer: Vulkanismus und die Naturgeschichte der Insel Santorin*, 1994
English edition first published 2000

Printed in the United Kingdom at the University Press, Cambridge

Typeface Scala 10.25/15pt. QuarkXPress [WV]

A catalogue record for this book is available from the British Library

Library of Congress Cataloguing in Publication data
Friedrich, Walter L. (Walter Ludwig), 1938–
Fire in the sea : The Santorini volcano: Natural history and the legend of Atlantis/
Walter L. Friedrich; translated into English by Alexander R.
McBirney.
p. cm
Includes bibliographical references and index.
ISBN 0 521 65290 1 (hardbound)
1. Santorini Volcano (Greece)–Eruptions. 2. Geology – Greece –
Thera Island Region. 3.Volcanism – Greece – Thera Island Region.
4. Minoans – Greece – Crete. 5. Civilization, Aegean. I. Title.
QE523.S27F65 1999
551.21'09495'85–dc21 99–25753 CIP

ISBN 0 521 65290 1 hardback

To my sons, Andreas and Michael, and my grandchildren, Anna and Jakob.

Contents

Preface

Jules Verne, in his famous science fiction book, *20,000 Leagues Under the Sea*, described how the submarine vessel *Nautilus* surfaced in the hot water around Santorini volcano. Captain Nemo and his crew were astonished to see the spectacular 1866 eruption of Georgios volcano on Nea Kameni. Today, the volcano is dormant, but it could return to life at any time. Tourists that walk over its black lava fields in the scorching summer heat have only the weak emissions of hot steam in the crater to remind them that Georgios is still alive. Even today one can explore the inner part of the main volcano of the Santorini very much in the way Captain Nemo did by sailing into the bay and admiring the multicolored inner walls of the gigantic volcanic caldron where an extraordinary part of the earth's history is revealed.

It is not just earth scientists who are irresistibly drawn to this volcanic island; it holds an equal fascination for archaeologists as well. Beneath the thick shroud of white pumice that mantles the rim and most of the outer slopes of the volcano lie the remains of an advanced culture – a Bronze Age Pompeii that, after more than 3600 years, is still being uncovered in the excavations at Akrotiri. This 'Fantastic Island' will enchant anyone who appreciates the natural world and the history of early civilizations. If it were up to me, I would reintroduce the name the island had in antiquity: Calliste – the most beautiful.

Acknowledgements

Fire in the Sea would never have been published without the joint efforts of several people. Two of them, however, had a special part in that. Alexander McBirney, Emeritus Professor of the University of Oregon kindly translated and improved my book through his enormous knowledge and inspiring discussions on Santorini and in Aarhus. Karl Pedersen, Dean of the Faculty of Science at Aarhus University, through his inspiring enthusiasm, helped me to provide financial support from the Faculty and the State Science Foundation. I should like to express my warmest thanks to them both.

My interest in the natural history of Santorini was provoked by Professor Martin Schwarzbach at Cologne and by Professor Hans Pichler at Tübingen. I am also greatly indebted to the latter for introducing me to the complex volcanic history of Santorini. The field trips to Santorini that I have carried out since 1975 have been financially supported by the Carlsberg Foundation in Copenhagen. I also wish to thank my Greek colleagues, especially the Geological Service of Athens (IGME) and the Greek archaeologists Professor Christos G. Doumas and Dr. Mariza Marthari, engaged in the excavations at Akrotiri for their permission to let me study the excavations and for numerous valuable discussions. I am also endebted to Tim Druitt and Floyd McCoy for valuable information they placed at my disposal. Grethe Nielsen, Ulla Viskum, Allan Kragelund, Birgit Kragelund and Mette Dybdahl provided invaluable technical assistance in preparing the manuscript and illustrations.

I should also like to thank my family, friends, colleagues, and students who helped me in so many ways. On our many field trips they assisted with my investigations and helped collect information of many kinds. Among these are Kirsten Søholm, Kirsten Gomard, Rud Friborg, Alfred Böttcher, Evangelos Velitzelos, David Dilcher, Horst Noll, Ole Bjørslev Nielsen, Hans Dieter Zimmermann, Marit-Solveig Seidenkrantz, Karen Luise Knudsen, Sidsel Grundvig, Ulrike Eriksen, Birgitta and Erik Hallager, Asger Ken Pedersen, Erik Schou Jensen, Svend Erik Rasmussen, Henrik Tauber, Bjørn Buchardt, Lotte Melchior, Sigrid

Hedegaard, Anne Clemmensen, Mikael Lund, Karin Knudsen, and Tom Pheiffer.

It was a special pleasure when the late professor, Arne Noe-Nygaard, joined us on two of our visits to Santorini. I would also like to thank Peter M. Nomikos for a wonderful trip to the Christiana Islands on board of Northwind II and Calliste. I shall always be grateful to Katina Gavala, Parthenios Gavalas, Natasha Koroneoy, Hannelore Sigala, Jutta Sigala, Charalambos Sigalas, Stellios Menetidis, Susanne Aarburg and Father Nicolas Kokkalakis for their warm hospitality and kind assistance on Santorini. Finally, I should like to thank Matt Lloyd, Tamsin van Essen, Stephanie Thelwell, Sue Tuck, and Alison Litherland, all from CUP, for their craftsmanship in producing the book.

Introduction

Volcanoes leave their mark on the history of the earth through their direct or indirect impact on nature and humans. One need only consider the effects of recent eruptions, such as those of Unzen (at Kyushu, Japan, 1990–93 eruptions) and Pinatubo (at Luzon, Philippines, 1991 and 1995 eruptions). Why are volcanoes concentrated in certain regions? What influence have they had on the climate, glaciation, and vegetation of the earth? These questions have been asked for centuries, and countless geologists, archaeologists, botanists, and climatologists have devoted much scientific research to finding answers.

This book deals with the historical development of one of the world's most remarkable active volcanoes, Santorini. Nowhere on earth can we learn so much about volcanism as we can on this island in the Greek Aegean Sea. We find there material evidence that enables us to trace the volcano's activity for nearly two million years and to construct a picture of the devastation it wrought over the course of time. One of the highlights of this history is the Minoan eruption, probably the greatest volcanic catastrophe of the Bronze Age. The event had a severe impact on much of the western world and contributed to the decline of the Minoan civilization. Scholars continue to discuss some of its possible effects. For example, the question of whether the sudden darkening of the sky mentioned in the *Argonautica* – one of the oldest Greek Myths told by Apollonius of Rhodes – could be a reminiscence of this eruption. It has also been warmly debated whether the darkness in the Bible – one of seven plagues of Egypt – could be attributed to this eruption. Similar discussions deal with the question of whether Santorini was the fabled island of Atlantis described by Plato. Studies of Santorini show that the event had such an impact on the people of that time that the memory of it has been passed down in written accounts or as legends.

Santorini has an unusual geological situation in a zone of intense deformation between the converging continents of Europe and Africa. Like the other Aegean volcanoes that are adding new material to the earth's crust, Santorini is a product of the collision of the two continents. Studies of these Aegean volcanic islands and the natural events that have occurred on them have contributed to the development of a number of basic geological concepts. One of the most famous of these was the theory of 'craters of elevation' which, when it was proposed by Leopold von Buch early in the nineteenth century, stimulated heated discussions, especially among scientists familiar with the geology of Santorini. According to von Buch's theory, large craters, such as the bay of Santorini, were the result of the expansion of a volcano by gas and molten magma; the swelling eventually opened fissures and triggered an eruption. Virtually all the noted naturalists of the day, including Alexander von Humboldt and Charles Lyell, contributed to the discussions, but it was the detailed geological studies of Santorini by Ferdinand Fouqué (1879) that finally put the theory to rest.

Santorini again attracted worldwide attention when Spyridon Marinatos (1939) proposed that the Bronze Age eruption of Santorini was responsible for the sudden demise of the Minoan civilization of Crete. Though received at first with skepticism, Marinatos' hypothesis was viewed with much greater interest when extensive layers of young pumice were found on distant islands and on the floor of the eastern Mediterranean. Although dating later showed that the Minoan eruption occurred well before the decline of the civilization on Crete, the controversy triggered a surge of intense investigations by geologists and archaeologists from many nations.

Today, most geological studies of the island are centered on the mechanisms of volcanic eruptions and the relationship of volcanism to large-scale tectonic features in the Mediterranean. Few regions have been the subjects of such intensive study. The great quantities of published work describing this research continue to increase. The wealth of information on Santorini resembles a fractal: no matter which field of science is explored each piece of new knowledge leads to new insights on another level. For natural scientists and archaeologists it has been a veritable Eldorado where specialists in different fields can work in close cooperation. In that sense, Santorini is a natural experiment in which almost all the natural sciences play a role: geophysicists determine the structure of the earth's crust and lithosphere; petrologists use the chemical composition of the rocks to define the origin of magmas; and paleontologists find the ages and ecological conditions in which the flora and fauna lived in the past. Thus, Santorini offers the natural scientist many possible windows through which we can gain an insight into the earth's past.

The landscape of Santorini is uniquely fascinating. It resembles a giant circus arena surrounded by circular stands. To geologists, the bay is known as a caldera, the Spanish word for a cauldron or kettle. We now know that even in the Bronze Age the islands of Thera, Therasia, and Aspronisi formed a single ring and from the settlements on its rim the inhabitants had a splendid view of a panorama that differed little from that of today.

Santorini, which in ancient times was called Calliste, found itself in the center of a region of old cultures. Lying between the Greek mainland, Crete, Asia Minor, and Egypt, it could quickly pick up and assimilate cultural changes from its neighbors. And in turn, it served as a crossroads where knowledge or innovations developed in one place were quickly disseminated to the surrounding region. Take, for example, the appearance of the island known as Hiera in the second century BC. The event was well recorded by the scholars of that time, and from their descriptions we can easily determine when it occurred. Strabo (66 BC to AD 24) tells us in his work *Geography* (1.3.16): 'For midway between Thera and Therasia fires broke forth from the sea and continued for four days, so that the whole sea boiled and blazed, and an island was gradually elevated, as though by levers. The island was a burning mass with a circumference of twelve stadia. When the eruption came to an end, the first people to venture onto the scene were Rhodians who, at that time had maritime supremacy. They erected on the island a temple in honour of Poseidon Asphalios.' We also have quite similar accounts of the volcanic events at Santorini from other authors. Volcanic events in this region were mentioned by Pindar, Herodotus, Callimachus, Apollonius of Rhodes, Seneca, Pliny the Elder, Orosius, Dio Cassius, Plutarch, Pausanias, Justin, Eusebius, and Ammianus Marcellinus – all well-known scholars closely associated with the cultural history of the Old World. The appearance of a new island was for the people of that time, as it is even today, such a remarkable, if not divine, event that it was well recorded. The Greek names Hiera, the 'holy', and Thia, the 'Godly', that were given to the new islands are good examples of this. The same fascination that such a natural event arouses even today is found in the writing of Seneca, on the formation of islands in the Aegean, when he asked how it was possible that fire in the sea is not extinguished, even when it occurs under a great mass of water. 'Marvels of the sea-fairer' is another account, in the work of Justin (Justinus *Trogi Pompeii*, 30.4.4), who lived in the second century AD and certainly reported on the origin of Hiera.

A further example is the legend of Atlantis, which Plato gave us in the dialogues 'Kritias' (Critias) and 'Timaios' (Timaeus). He described the sinking of an island that had a flourishing culture. Some scholars believe that at the core of this legend one can recognize the Minoan eruption that inflicted a very similar fate upon Santorini.

Interpretations of the rise of the island complex from the sea changed over the course of time. In ancient times the Gods were credited for all forces such as this that governed the processes of our earth. For example, Poseidon was said to control the seas and earthquakes. Thus, when the eruption of AD 46 occurred on the night of the eighth-century festival of Rome, it was seen as a clear expression of the Gods' will. At that same time there occurred a conspicuous total eclipse of the moon, and the firebird Phoenix of Arabia (a comet?) appeared to proclaim the birth of a new island in the Aegean. As Aurelius Victor reported in his fourth-century work, *Historiae Abbreviatae*, this series of evil omens could have only one meaning: the decline of the Roman Empire was imminent.

In the Christian era, devastating volcanic eruptions were interpreted as signs of God's wrath over the sins of man. Fateful importance was assigned, for example, to the eruption of AD 726 on Palea Kameni, which gave impetus to the crisis between Rome and Constantinople. Today, however, people look to science; they believe in specialists' forecasts of impending eruptions and many trust blindly the ability of volcanologists to inform them correctly when to anticipate a catastrophe. If divine forces are involved, they are no longer a factor in these forecasts.

Using the example of Santorini, this book traces the development of an active volcanic island. Special importance will be assigned to the timing and variety of the volcanic events that have taken place since the Late Pliocene, a period of about two million years.

We are reminded that the forces responsible for the volcano's activity are still operating today. We shall also examine paleontological discoveries as well as traces of early human settlement, but the main objective is to analyze the Minoan eruption that buried the flourishing Bronze Age settlements on Santorini under a great shroud of volcanic ash and left for posterity a prehistoric Pompeii.

Part 1

The geological framework

"The same thing [an eruption between Thera and Therasia] happened again in our own time during the second consulship of Valerius Asiaticus. Why do I mention these curiosities? So it may be clear that fire is not extinguished even by a sea covering it and that its force is not prevented from bursting out even by an enormous mass of water. Asclepiodotus, the pupil of Posidonius, says that the height of water through which the fire reached before it had broken through was two hundred feet."

Seneca, *Naturales Questiones*, 2.26.6

Seneca lived in the first century of the Christian era. He was the palace philosopher of the emperor Nero in Rome.

1

The geography of Santorini

In recent centuries the Aegean volcanic islands of Santorini have repeatedly changed not only their appearance but their names as well. The ancient Greek called Santorini Calliste – the most beautiful.

The group of islands known as Santorini belongs to the Cycladean Archipelago, which lies in the southern Aegean and extends through numerous islands and islets (Fig. 1.1). They are about 120 kilometers north of Crete, and in keeping with the meaning of their former name in ancient Greek, Calliste, are the most beautiful of all the Greek islands. Santorini was originally a single, more-or-less round volcanic island with several volcanoes, which draped their ash, slag, and lava flows over an older core of non-volcanic rocks. Before the catastrophic Minoan eruption of the Bronze Age gave the volcanic edifice its present form, it was still a connected island ring. Geologists used a similar designation, Stronghyle, meaning 'round' in Greek, in the last century to refer to the island complex. According to Galanopoulos and Bacon (1969), the name must have been passed down from ancient times. That is certainly reasonable, for the term was used to designate Thera (or Santorini) by Pliny, who lived from 24 to 79 AD, and, of course, witnessed the eruption of Vesuvius in 79 AD. Pliny (*Naturalis Historiae I.III.94*) says: 'Between here and Sicily lies another (island), once called Therasia but now Hiera, because its volcano is sacred. Flames come from the mountain every night. The third is Stronghyle, six leagues from Lipari in the direction of the rising sun where the Aeolos rules . . . .' Indeed, Pliny spoke of Therasia and Hiera as part of the first group, but today presumably one of the modern volcanic islands belonging to the Eolean Islands and not the similarly named island of the Santorini group. The name Stronghyle (pronounced Strongili) is also used for a

Figure 1.0 The town of Oia is situated in the northern part of Thera, the main island of the Santorini group. Standing near the top of the caldera wall, it is reached from the sea by climbing the stairway visible in this view or by coming overland from the harbor at Athinios. The earthquake of 9 July 1956 destroyed Oia, but most of the town has now been rebuilt.

volcano of the Eolean islands, namely Stromboli, for which I give further details in Chapter 12.

In the Bronze Age a great eruption devastated the island so that today there are only a few remains left that are grouped around a large bay. The large central basin or caldera is divided into the three remaining parts of the older ring-island, Thera, Therasia, and Aspronisi. In the middle of the caldera are two other volcanic islands, called Palea Kameni and Nea Kameni. They are certainly younger than the other islands, for they were formed in historic times and are today still active (Fig. 1.2).

Although the surface area of the main island of Thera is only 76.2 square kilometers (Table 1.1) and the population only 9360, every year the island is visited by thousands of tourists from many nations. The main attraction of the island is the excavated Bronze Age settlement of Akrotiri, which, during the height of the season, has as many as 2000 visitors daily.

The climate of Santorini is typical of the Mediterranean zone with winter rains, quite dry summers, and cool winters. In the summer, northerly winds, Etesian (Greek *meltemi*), prevail in the eastern Mediterranean region and bring welcomed cool air to relieve the scorching summer heat. The sun shines for a total of about 3250 hours a year. Storms and violent winds mark the winter months and the higher elevations may even receive snow (Fig. 1.3).

Thera – the main island of Santorini

Thera (see Fig. 3.9) is the largest remnant of the original ring-island. The core of the crescent-shaped island has the oldest rocks of the island group. Originally laid down as sediments about 200 million years ago, these rocks have been metamorphosed, deformed, and uplifted so that they now form the highest hill on Santorini, Profitis Elias, which rises to an elevation of 565 meters above the sea. These ancient metamorphic rocks extend westward along the ridge at Pirgos and as far as the caldera wall at Athinios and Cape Plaka.

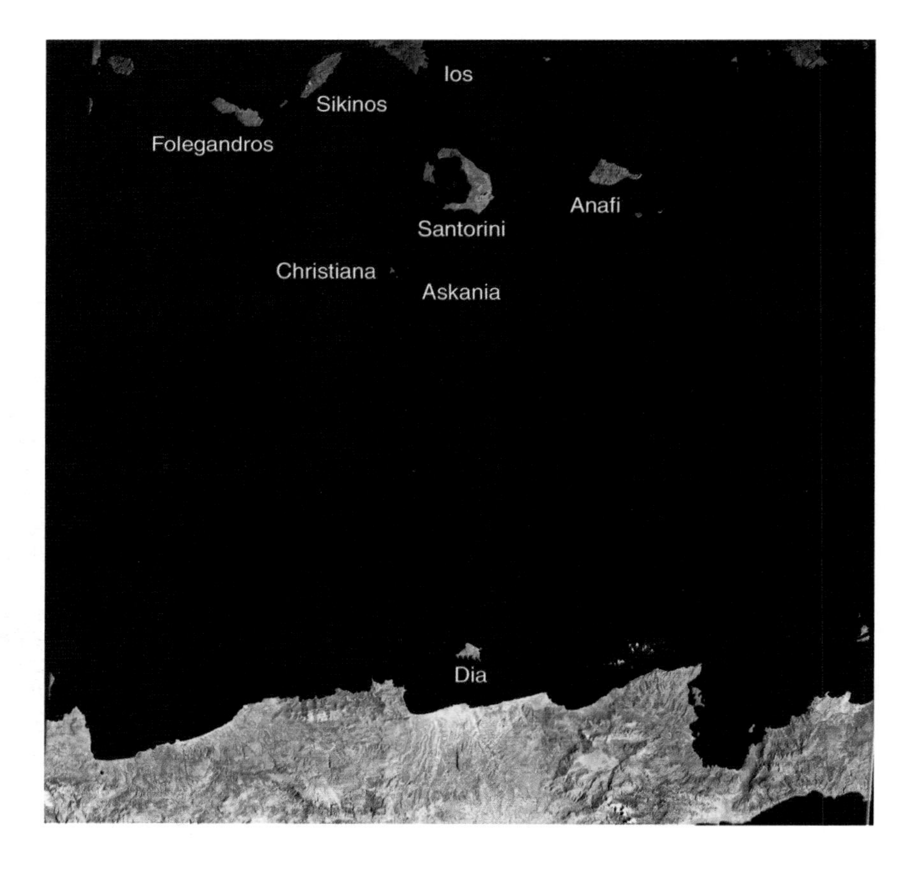

Figure 1.1 This satellite image taken on 6 September 1977 shows Crete (below), Santorini (near the upper center), and other islands in the Aegean Sea. The distance between Crete and Santorini is about 120 kilometers. Photo from Telespazio ESA – Earthnet, September 1977.

They also form the cliffs of Platinamos Ridge, which even now has a few windmills on its crest. In the east near the airport, a small knob of limestone rocks, appropriately named Monolithos (Greek *monos* for 'single', *lithos* for 'rock') projects upward about 20 meters. It forms an island rising through the thick mantle of pumice that the Minoan eruption laid down over most of Thera.

Apart from these metamorphosed relics of the pre-volcanic period of Santorini, the island consists entirely of volcanic rocks. In the northern part of Thera they make up the volcanic cones of Megalo Vouno (330 meters), Kokkino Vouno (283 meters), and Mikro Profitis Elias (314 meters), all of which are built of lava, ash, and slag.

Fira (or Phira), the main town of the island, stands on the crest of the caldera rim, where its picturesque, white houses and churches look out over the bay. From the small harbor at Katofira it is reached by a long, winding stairway and, in recent years, by a funicular. The main port where most large ships anchor lies farther south at Athinios (Fig. 1.4). To the north lie the villages of Oia, also called Apanomeria (Fig. 1.0), and Phoenikia. In the middle of the island are the smaller towns of Karterados and Mesaria, Pirgos, Vothonas, Exo Gonia, Megalochori, and Emporio. On the eastern shore of the island are the villages of Kamari and Perissa, and in the south the village of Akrotiri. Since about 1977 there has also been an airfield on Thera, in the vicinity of Monolithos.

Figure 1.2 The flooded caldera of Santorini has an area of about 84 square kilometers. Near its center are two volcanic islands, Palea Kameni (on the left) and Nea Kameni (right).

Table 1.1 *Santorini (Thera and Therasia) compared with other bigger islands of the Cyclades (from National Statistical Service of Greece 1994, 1995)*

Island	Area (km²)	Population	Inhabitants (per km²)	Height above sea level
Naxos	389.4	14838	38.1	1004
Andros	383.0	8781	22.0	994
Paros	196.7	9591	48.7	771
Tinos	197.0	7747	39.3	713
Milos	158.4	4390	27.7	751
Kea	131.7	1787	13.5	570
Amorgos	121.5	1630	13.4	822
Ios	108.7	1654	15.2	732
Kythnos	99.4	1632	16.4	368
Mykonos	86.1	6170	71.7	392
Syros	84.1	19870	236.3	442
Thera (Thira)	76.2	9360	122.8	566
Therasia (Thirassia)	9.2	233	25.3	295
Serifos	74.3	1095	14.7	586
Sifnos	77.4	1960	25.3	680

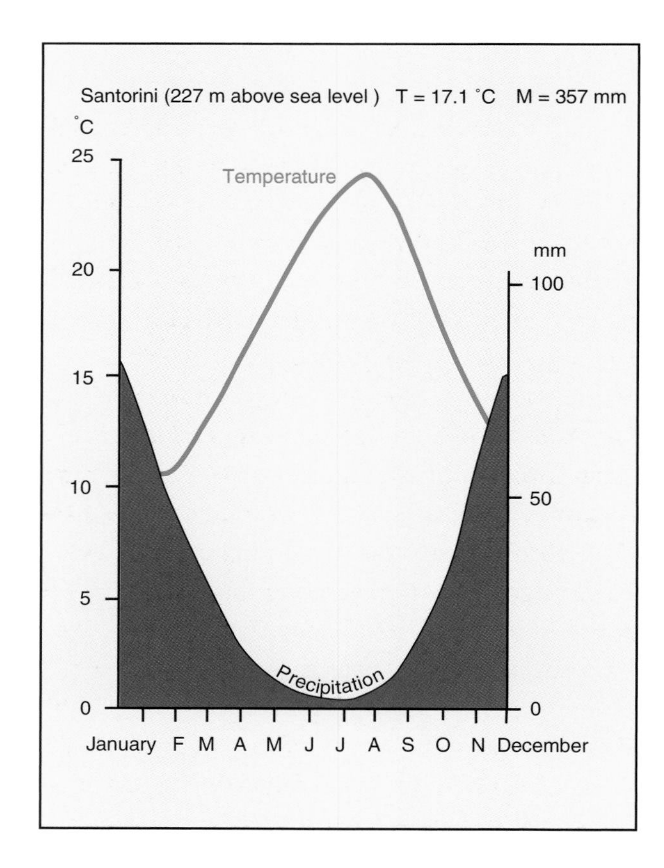

Figure 1.3 Santorini has exceptionally warm summers and stormy, humid winters. The diagram above shows average monthly temperatures (red curve) and precipitation (blue). The mean annual temperature (T) is 17.1 °C and the mean maximum precipitation (M) is 357 mm. The height of the measuring station on Thera is 227 meters above sea level. Modified after Walter and Lieth (1967).

Therasia – the Little Thera

The island of Therasia, the Little Thera, which already had this name in ancient times, forms the northwestern part of the ring of islands. It consists entirely of volcanic rocks, which at Viglos Vouno reach the height of 295 meters. The largest town is Manolas, which stands on the caldera rim above the harbor of Corfos. Manolas is reached by a steep, winding stairway from the harbor and by a new road from Riva, the landing place in the northwestern part of the island. The village of Potamos, not far from Manolas, is situated in a small valley, which during the winter rains has a stream (potamos in Greek). Until the earthquake of 1956 there was still the village of Agrila, but

Figure 1.4 The capital city of Santorini is Fira, visible in the upper part of the photograph. It is situated about 220 meters above sea level close to the rim of the caldera. For many years it could be reached only on foot or donkey using the steep stairway, but a cable car now makes the trip much easier. The small harbor below the town is accessible for boats and small ships. Larger vessels use a buoy several hundred meters off shore. Owing to the great depth of the caldera they cannot use anchors. Passengers going ashore are transferred to small boats. A larger harbor located at Athinios can be seen in the lower part of the picture.

today it is gone. The monastery of Kimisi on the southern tip of Therasia is occupied for only a few days in August for the feast of the Blessed Virgin.

Aspronisi – the White Island

As the Greek name Aspronisi (*aspros* – 'white', *nisos* – 'island') appropriately expresses it, this small island is covered with a thick layer of white pumice laid down by the Minoan eruption. Aspronisi is uninhabited and, even though it is only 60 meters at its highest point, its interior is almost inaccessible.

The Kameni Islands

The island known as Palea Kameni, meaning 'Old Burned' in Greek, is well named. It was the first to be formed after the Minoan eruption. From various written sources we know that volcanic activity started here around 197 BC (Chapter 12). Except for a herd of a few goats, Palea Kameni is now uninhabited. Its truncated form indicates that, in the past, it was substantially larger. This is shown most clearly by the steep cliff on its northwestern side. If one compares the island with Nea Kameni, a nearby island of similar size, one gains the impression that the present Palea Kameni is only a small part of what was once a much larger island. The largest part probably disappeared into the depths of the caldera during the early centuries of the Christian Era or the early Middle Ages. Moreover, the vegetation, which varies with the type of lava and the amount of weathering and erosion, shows that the island consists of parts with differing ages. The oldest part probably grew during two eruptive episodes: the first phase produced a lava dome in the central region that rises to a height of 103 meters and the second a tongue of lava in the northeast (Fig. 1.5). A steep cliff that rises from the sea cuts off the older part on the southeastern side. The surface of the main part of the island is cut by gaping fractures that have opened par-

Figure 1.5 Palea Kameni is one of the two young islands within the caldera of Santorini. It has two distinct parts. The older part terminates in a steep escarpment along its northeastern shore. The cliff reveals a section through a lava dome that has the form of a gigantic onion. The layered parts of the dome were caused by shrinkage during cooling of the highly viscous magma. The younger part of the island is a fresh tongue of lava that forms the northeastern tip of the island.

allel to the face of the cliff. The younger part is easy to recognize because it has a quite different appearance. It is flatter and consists of block lava made up of shiny black glass. This flat lava field on the northeastern side, near the church of Agios Nikolaos and at a small inlet for boats, probably dates from the eruption of 726.

The 'New Burned' Island (Nea Kameni) is the youngest volcanic island in the caldera. It was formed in 1707 by the coalescence of newly erupted lava with the earlier rocks of Mikra Kameni. Its almost circular, uniform shield is made up of lava, slag, and ash. The summit crater, Georgios, is the highest point on the island (124 meters). In its walls certain spots (solfatara and fumaroles) still give off warm sulphurous gases.

The Caldera – the large central basin

The ring of islands formed by Thera, Therasia, and Aspronisi encloses a marine basin, or caldera, having a surface area of 84.5 square kilometers. Bathymetry shows that the caldera boundaries are not uniformly shaped, especially in the four sub-basins. The greatest depth is in the northeastern basin, which reaches nearly 400 meters below sea level. This basin may have been formed during the Minoan eruption while the other three were formed earlier (see Chapter 11). If one could imagine the caldera without water, it would reach a depth of 700 meters below the volcano of Megalo Vouno.

Kolumbo – the invisible volcano

Not all parts of the Santorini volcanic edifice are visible today; major parts are hidden under the surface of the sea. For example, the volcano Kolumbo, also called Columbos, which rose from the sea during its period of activity in 1649–1650 (see Chapter 12), is now totally submerged. Vougioukalakis *et al.* (1994), who recently studied the volcano with the help of divers, have been able to obtain much new information about this hidden volcano. Its name is probably derived from the Greek word *kolymbao* (= to swim) referring to the appearance and disappearance of the volcano that at times resembled a person swimming.

Situated 7 kilometers off the northeastern coast of Thera the cone measures 4 by 8 kilometers with its long axis oriented parallel to a northeast-trending tectonic structure known as the 'Kolumbo line'. Its highest point at the crater rim lies only 18 meters below sea level. A caldera is formed in the central part that reaches a depth of 512 meters below sea level.

The Christiana Islands – relatives of the Santorini volcanic group

The volcanic islands Christiani, Askania, and the reef Eschati (the name of which means 'the last one') are also part of the general volcanic complex of Santorini (Figs. 1.6 to 1.9). They lie at the southwestern end of a tectonic line running northeastward through the Kameni islands to the submarine volcano Kolumbo (see Chapter 12).

This line has affected the shape of the largest of the three islands, Christiani. This island is divided into two parts, a higher part in the northeast and a lower part in the southwest. Erosion of the elevated part by the prolonged effects of wind and rain has stripped the surface bare and deposited a thick mantle of debris on the surface of the lower part. At present we have no radiometric data to establish the age of these rocks, but the accumulation of debris indicates an age of several hundred thousand years, similar to that of the Akrotiri peninsula on Thera.

At present the Christiana Islands are not inhabited, but ruins, road-like structures for collecting water, and other signs of human activity, including traces of Bronze Age and medieval settlements, clearly show that for many centuries they were densely populated. A small church on Christiani is still visited by fishermen and hare-hunters coming to the island.

Santorini, an island with deep roots in history

Few islands in Greece are mentioned as often in ancient literature and myths as Santorini. Even in the oldest written accounts one finds discussions of this interesting group of islands, which in the course of time has repeatedly changed not only its appearance

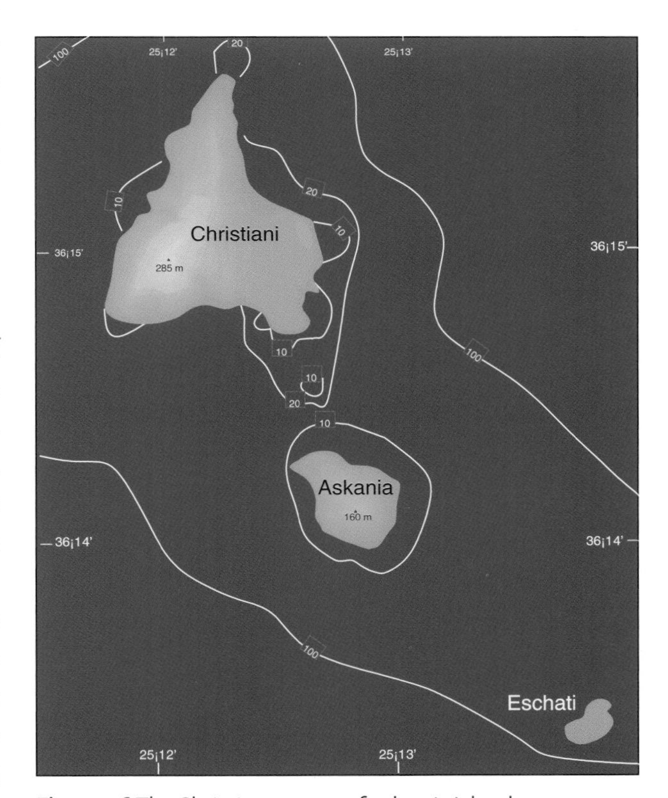

Figure 1.6 The Christiana group of volcanic islands – Christiani, Askania and the small reef Eschati – are situated near the southwestern end of a submarine ridge that extends northeastward through Santorini to the submerged volcano Kolumbo.

but also its name. Whenever a dramatic event, such as a volcanic eruption or the appearance of a new island, has occurred in this region it has left a deep impression on the inhabitants, who have passed down written or oral accounts to their descendants. It is not always clear whether the names in these accounts refer to the entire archipelago or only to an individual island. Since the Bronze Age, people from Crete and the mainland have colonized Santorini. The archaeological record of these people goes back to the thirteenth century BC. Later, it was the Phoenicians, Spartans, Ptolemeans, Romans, Byzantines, Franks, and Turks who left their marks on the history, language, and architecture of the island. The list of names that have been used for Santorini includes Calliste, Stronghyle, Thera, and perhaps even Plato's so-called Atlantis, the mythical island created by Poseidon (Fig. 1.10). After the great pumice eruption of 1640 BC the islands in the caldera were given names, such as Hiera, Thia, Palea, Mikra, and Nea Kameni. The earliest mention of the name Thera is in verse 10 of the fourth Pythian Ode of Pindar (522–441 BC), and the name Calliste is used in verse 258. In this ode, Pindar reports the voyage of the argonauts and their discovery of the island of Thera, off the coast of Libya. Kallimachos, who lived around 250 BC, spoke of the same events, but only fragments of his poetry have survived. Apollonius Rhodius, who lived in the third century BC, also referred to the same story of the *Argonautica*, in which the gods often played a prominent part. Poseidon and his family are named in

Figure 1.7 The three islands of the Christiana volcanic group are shown in this photograph. Christiani is in the foreground, Askania in the middle, and the small reef of Eschati in the background and to the left of Askania. Although uninhabited today, numerous traces show that they were populated from prehistoric times until the 1970s.

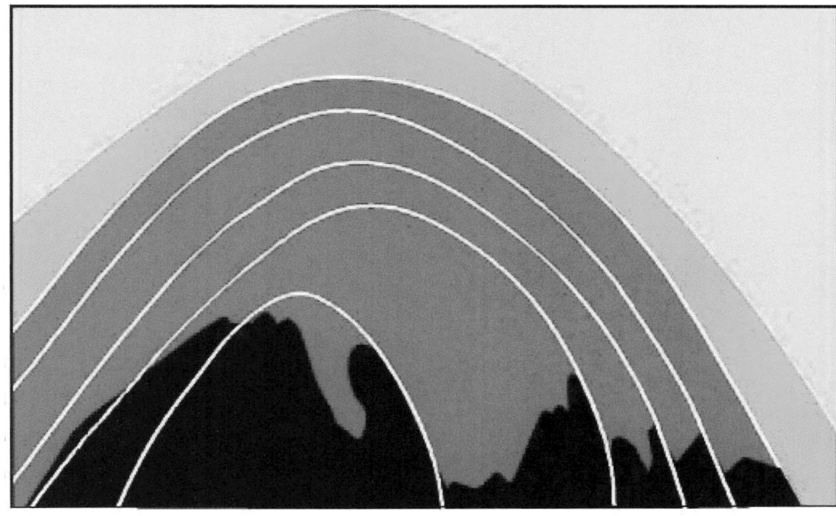

Figure 1.8 (left) The reef of Eschati has a height of about 15 meters (top). It consists of volcanic rocks (basaltic andesite) that intruded and elevated the sediments of the seafloor. Most of the sediments are now eroded away leaving only the hard core of the updomed area (bottom). Yellow: marine sediments; red: volcanic intrusion.

Figure 1.9 (below) The island of Askania resembles the updomed part of Akrotiri peninsula in that it is a volcano that started under water and emerged above sea level. Fossiliferous marine sediments are found in the lower part and columnar lavas in the upper part.

conjunction with Thera in several of the old accounts. On their homeward journey, Euphemos, the son of Poseidon, received a clod of white soil as a hospitality gift from Triton. After leaving Anaphe, he had a warning sign in the form of a clap of thunder and threw the soil into the sea. From this was formed the island of Calliste, the most beautiful, which was later called Thera (Apollonius Rhodius *Argonautica,* 4: 1750ff, 1730ff). The name Thera comes from the commander Theras, son of Autesion of Sparta, who, as Herodotus of Halicarnassus (484–424 BC) tells us, had earlier led colonists to Calliste (see Box 1.1). From this text we also learn that Kadmos, who escaped from Zeus and was looking for Europa, provided a link to one of the oldest of all Greek legends: Agenor, the son of Lybia and Poseidon, left Egypt and settled in the land of Kanaa where he married Telephassa. This union brought him five sons (Kadmos, Phoenix, Cilix, Thasus, and Phineus) and one daughter, Europa. Zeus became devoted to Europa and, changing him-

Figure 1.10 In the ancient town of Thera on the low hill of Mesa Vouno one can see the sacred site of Artemidoros. It consists of altars, inscriptions, and bas-relief figures erected around 280 BC by Artemidoros, the admiral of the Ptolemean fleet. Beside his portrait, a lion, a dolphin, and an eagle symbolically represent Apollo, Poseidon, and Zeus. Poseidon is also mentioned prominently in literature describing the origins of Calliste and Thera. Herodotus tells us that Cadmus visited Thera when he was searching for Europa and, in 197 BC, the people of Rhodes honored Poseidon by building a temple on the new island of Hiera. In addition, Plato mentions Poseidon in connection with the legend of Atlantis that some scholars associate with Santorini. The dolphin has a length of about one meter and above it is an inscription chiseled into the rocks saying 'To Poseidon, God of the sea, Artemidoros has chiseled in the everlasting rock a dolphin, considered friendly to people, to honor the gods.' Drawing from Hiller von Gaertringen (1902). Around the portrait of Artemidoros is a band of text which says (Hiller von Gaertringen, 1904, p. 97): 'As a memorial for Thera, the name of Artemidoros will not vanish, as long as stars on heaven and ground of the earth exist.'

Box 1.1

The oldest mention of the name Thera is found in the Pythian, Odes of Pindar, who lived from 522 to 441 BC.

. . . Battus the colonizer of fruitful Libya, in order that he might at once leave the holy island [Thera], and build, on a gleaming hill, a city of noble chariots, and thus, in the seventeenth generation, fulfil the word spoken at Thera by Medea . . .

Pindar (*IV.5–10*)

In the same ode the name Calliste, the most beautiful, is also mentioned.

There it was that the race of Ephemus was planted, to increase for ever in the days to come; and, having mingled with the homes of the Lacedaemonians, in due time they went and dwelt in the isle once called Calliste.

Pindar (*IV.258*)
(From the English translation by Sir John Snadys)

The islands near Crete are Thera, the metropolis of the Cyrenaeans, a colony of the Lacedaemonians, and, near Thera, Anaphe, where is the temple of the Aegletan Apollo.

Strabo (*Geography 10.5.1*)

There were in the island now called Thera, but then Calliste, descendants of Menbliorus the son of Poeciles, a Phoenician; for Kadmos son of Agenor, in his search for Europa, had put in at the place now called Thera; and having put in, either because the land pleased him, or because for some other reason he desired so to do, he left in this island, among other Phoenicians, his own kinsman Membliarus. These dwelt in the island Calliste for eight generations before Theras came from Lacedaemon.

Herodotus (*IV.147*)

self into a bull, took Europa on his back and swam away. Europa bore Zeus three children: Minos, Rhadamanthys, and Sarpedon. After the disappearance of Europa, her father, Agenor, sent out his sons to search for her. He ordered them not to return without her. Not knowing in which direction the bull had traveled, they separated, each taking a different route. Kadmos, guided by his mother, Telephassa, sailed first to Rhodes and then to Thera, where he paid homage to Poseidon with a temple. The God Zeus is also connected with Thera (Santorini) in another tale, which tells us that the appearance of the island today is a result of a battle between Zeus and the Titans. As Zeus wanted to put an end to all the retreating Titans, he seized the core of Thera in his mighty hand, tore out its mid-section, and hurled it after them. Today one can recognize – at least according to Durazzo-Morosini (1936) – the traces left by his powerful hand. Four small inlets on the inner side of Thera (between Oia and the cliffs of Skaros, between there and Fira, between Fira and Athinios, and finally between Athinios and Balos), are said to correspond to the four fingers of his right hand. His thumb scraped a groove between Akrotiri and Therasia.

Vestiges of Santorini's earliest appearance in history, from which these legends and stories have been passed down, are still to be seen today in the ruins of the town of Ancient Thera on Mesa Vouno. From historical sources, most notably Herodotus, we know that the first settlers were the Phoenicians led by Kadmos (see Box 1.1). They lived on Calliste for eight generations. Then came Theras, the son of Autesion, with the Lacedemonians, who named the island Thera, after their leader Theras. After the ninth century Thera became a Dorian colony and an important site in the trade routes of the Mediterranean of that time. The Phoenician alphabet was introduced to the islands of Milos, Crete, and Thera at about the end of the ninth century. Archaeological evidence shows that during the Archaic Epoch, Thera had contacts with Crete and Paros, as well as the Greek Mainland and Rhodes. About 630 BC the Thereans founded Cyrene, which Herodotus tells us was their only colony in Libya (*IV.150ff*). The oracle of Delphi guided the leader of this colony, Battus. Thera's role in the trade of the Aegean area is shown by the fact that they had their own money. Coins from Thera dating from about 635 BC show two dolphins (Fig. 1.11) as their official emblem (Kraay 1966). During the classical

Figure 1.11 Two dolphins are shown on coins used on Thera around 635 BC. The back of the coin shows the sails of a windmill. Three times natural size. Photo from Kraay, 1966.

epoch of Greece (fourth to fifth century BC), Thera does not seem to have played an important role, because these coins disappeared during the Persian Wars. But after the decline of Athens' hegemony in the Aegean area, Therean money reappeared. During the Peloponnesian war the Thereans fought on the side of Sparta. In the Hellenic Epoch, Thera had an important place in the Ptolemean Empire. In addition, it had two sandy beaches that could be used by the fleet: Kamari and Perissa. The Romans, who left abundant signs of their occupation of Thera, also appreciated these advantages. Early Christian churches on Thera, such as the recently excavated one at the town Perissa, are also from Roman times. Thera had its own Christian bishop, Dioscouros (342–344), as early as the fourth century. Byzantine historians mentioned Thera in connection with the eruption on Palea Kameni in 726 (see Chapter 12). The German archaeologist Freiherr Friedrich Hiller von Gaertringen excavated and described the ruins of Ancient Thera, the city on Mesa Vouno, at the end of the nineteenth century (Fig. 1.12). In this town he found the shrine (Temenos) of Artemidoros, where we again encounter Zeus, Apollo, and Poseidon. Pindar, in the fourth Pythian Ode, mentioned these same deities, where he tells us about the naming of Thera. Carved in relief on the face of one of the rocks are an eagle, a lion, and a dolphin, which symbolize these gods. In numerous ancient writings the name Thera is found in different contexts, but mostly in connection with natural

Figure 1.12 The eagle (about 50 centimeters long) carved from the limestone at the sacred site of Artemidoros on Mesa Vouno symbolizes Zeus, father of the gods. According to Hiller von Gaertringen, the nearby inscription reads: 'To Zeus Olympios, Artemidoros has dedicated for all times an eagle, the high-flying messenger of Zeus, to the city and the immortal god.'

events, as will be described more fully in Chapter 12. Thus we note, for example, that according to Pliny the Elder, Thera was one of the important sources of the saffron that was so highly prized at that time.

The name Thera (or Thira) is now applied only to the main island, but it is also retained in the name Therasia. The name Santorin or Santorini, widely used today, originated at the time of the Crusades and the fall of Constantinople in 1204. At that time the conqueror of the archipelago, Marco Sanudo, gave Thera and Therasia as a fief to the Venetian Iacomo Barozzi, whose descendants ruled the islands until 1335. The Venetians had a landing place at Riva on the island of Therasia where they erected a chapel in honor of Santa Erene (Saint Irene), who had become a martyr in Thessalonika on the 23rd of March 303.

In 1335 Santorini became part of the 'Duchy of the Archipelago'. Documents from that period indicate that in 1336 the castle of Akrotiri was given over to the Gozzadini Family. In 1479, Santorini became the possession of Domenico Pisani; in 1480 it was given over to Giovanni III Crispo, and in 1492 was taken over by Venice. One finds the name Sant-Erini (or Santa Erini) on some old maps, including a 1421 map by the Florentine priest Buondelmonti. On his map of '*Santa Erini*' Bartolomeo dalli Sonetti depicted five castles and the Christiana Islands. The name 'La Christiana' is written in its correct place in the southwest of Santorini (Fig. 1.13). Furthermore, the name Sant Erini is mentioned in a poem by Bartolomeo dalli Sonetti in his book *Isolario* (from around 1485).

The Crispo Family owned Santorini until 1537, when the island group was lost to the Turks as a result of the third war between Turkey and Venice. In the period 1566–1579 Santorini belonged to the Jew Joseph Nazi, after which the Ottoman Rule continued until the Greek War of Independence in 1821. The war resulted in recognition of an independent Greek state in 1830. The Turks named the island Gozi, as we learn from Dapper (1730).

Today, the name Santorini is used more and more to designate the whole volcanic complex, which before the Minoan eruption had formed a complete island ring.

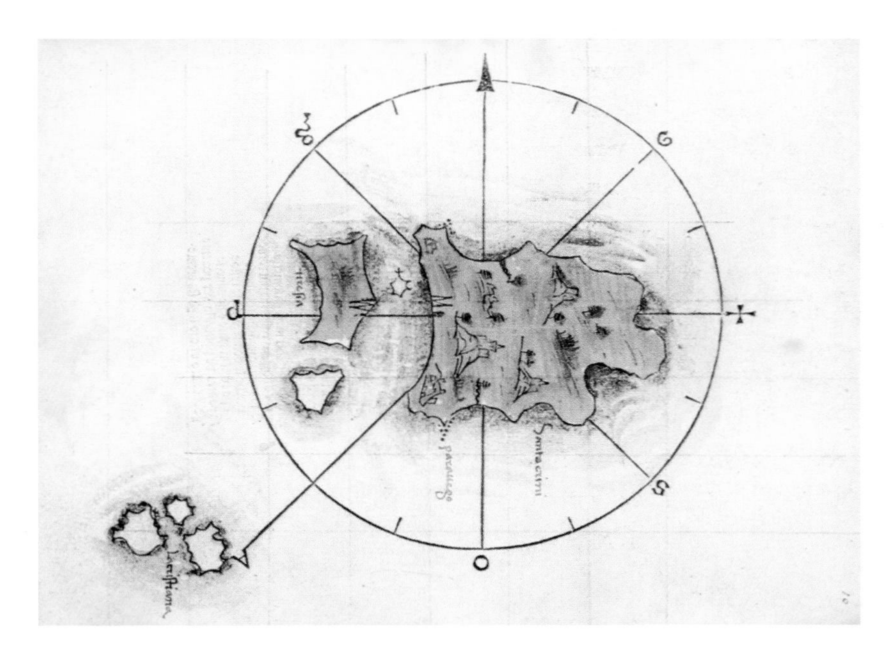

Figure 1.13 The map of Santorini by the Venetian Bartolomeo dalli Sonetti shows the name Santa Erini. The three Christiana islands (La Christiana) are seen in the southeast outside the circle. The colored woodcut can be dated before 1485, since it is dedicated to a person who died in 1485. From Monioudi-Gavala (1997).

2

Santorini and the puzzle of plate tectonics

The volcanic islands that form a curved line across the Aegean were produced by the convergence of two continental landmasses, Africa and Eurasia. This chain has been named the south Aegean volcanic arc or the Hellenic arc. Santorini is today the only active volcano in this arc; it has had three eruptions in the twentieth century alone.

In the year 1965, geology underwent a drastic change. New ideas suddenly replaced old deeply entrenched ones about the origins of continents and mountains, and a static view of the earth gave way to a dynamic one: plate tectonics, which soon became the fundamental concept of geology. Based on the theory of continental drift, first proposed by Alfred Wegener in 1915 and subsequently supported by new discoveries, it quickly gained wide acceptance among scientists in all realms of geology (Fig. 2.1). The most important lines of evidence in support of plate tectonics come from the magnetic stripes on the ocean floor and from abundant earthquakes that are observed along the margins of continents (Fig. 2.2). These were the keys to a new interpretation of our earth: earthquakes clearly indicate the relative motion between adjacent plates.

It has long been known that the direction of the earth's magnetic field is recorded in lavas when they cool and small crystals of iron oxides take on an orientation parallel to the lines of flux. We also know that the polarity of the magnetic field switches from time to time so that the north-seeking needle of a compass would point south. The last time this happened was about 700 000 years ago. In the 1960s, studies of the magnetic fields of the oceans showed that the sea floor had striped patterns in which the rocks had reversed polarities. It was soon noticed that the pattern of these stripes was symmetrical about a long ridge running

Figure 2.0 The caldera wall at Fira on the island of Thera reaches a height of about 220 meters. It represents a section of Santorini's volcanic history that spans nearly half a million years. At the very top are the white houses of Fira and below the town is the small harbor to which it is connected by a zigzag donkey path. The distinctive hill of Skaros can be seen on the upper left.

down the middle of the Atlantic Ocean (Figs. 2.1 and 2.2).

When it was recognized that the spacing of these stripes was proportional to the time scale of reversals of the earth's magnetic polarity, it suddenly became clear that the lavas on the sea floor were generated at the ridges and moving outward at a steady rate. Surveys of other oceans showed that this process was going on in all the oceans of the world, but the rates of spreading differed somewhat from one ocean to another. Looking at the Pacific Ocean, it was noticed that the stripes seemed to disappear when they reached the deep trench near island arcs and continental margins. It soon became apparent that the generation of new crust at the ocean ridges was balanced by consumption at the margins of the Pacific.

By far the majority of the world's earthquakes is concentrated along oceanic ridges and near deep trenches. When this distribution is viewed on a map of the globe, it defines the boundaries of large regions of crust that seem to behave like rigid plates that drift like sheets of ice on the surface of the sea. The motion associated with the earthquakes is divergent at the oceanic ridges where crust is being generated and convergent where it is being thrust back into the mantle at island arcs and continental margins.

Another discovery of special interest was made when geologists examined the volcanic regions of Japan, the Aleutians, the Andes, and Indonesia, where active volcanoes are arranged in chains along the margins of the oceans. This volcanism is associated with earthquakes concentrated along planes extending from the trenches down into the mantle. These zones

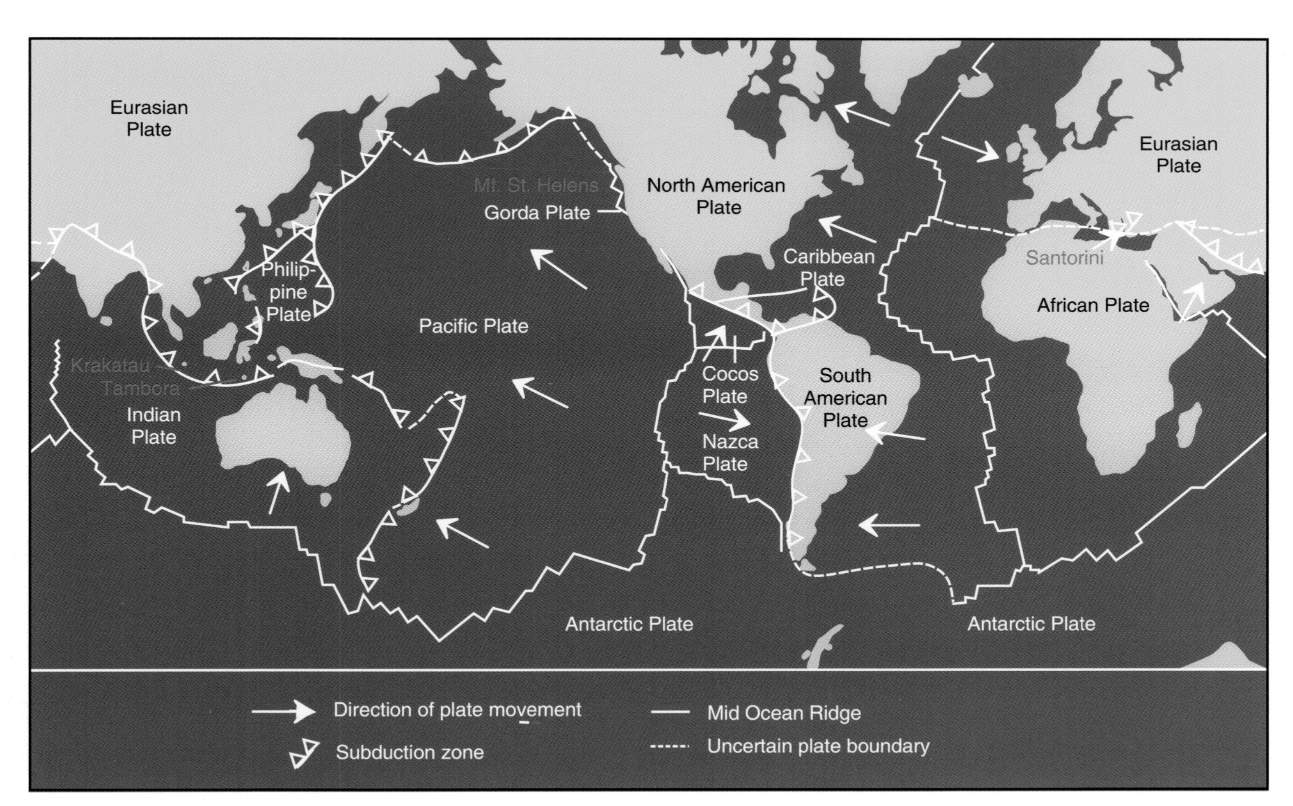

Figure 2.1 Our earth consists of gigantic plates, about 100 kilometers thick, that are constantly moving, like great sheets of ice floating on water. They approach each other, collide or move away. During a collision one of the plates may be subducted into the asthenosphere where it is consumed. Most of the earthquakes and volcanism on the earth are related to the movement of these plates. Modified after Pichler (1988).

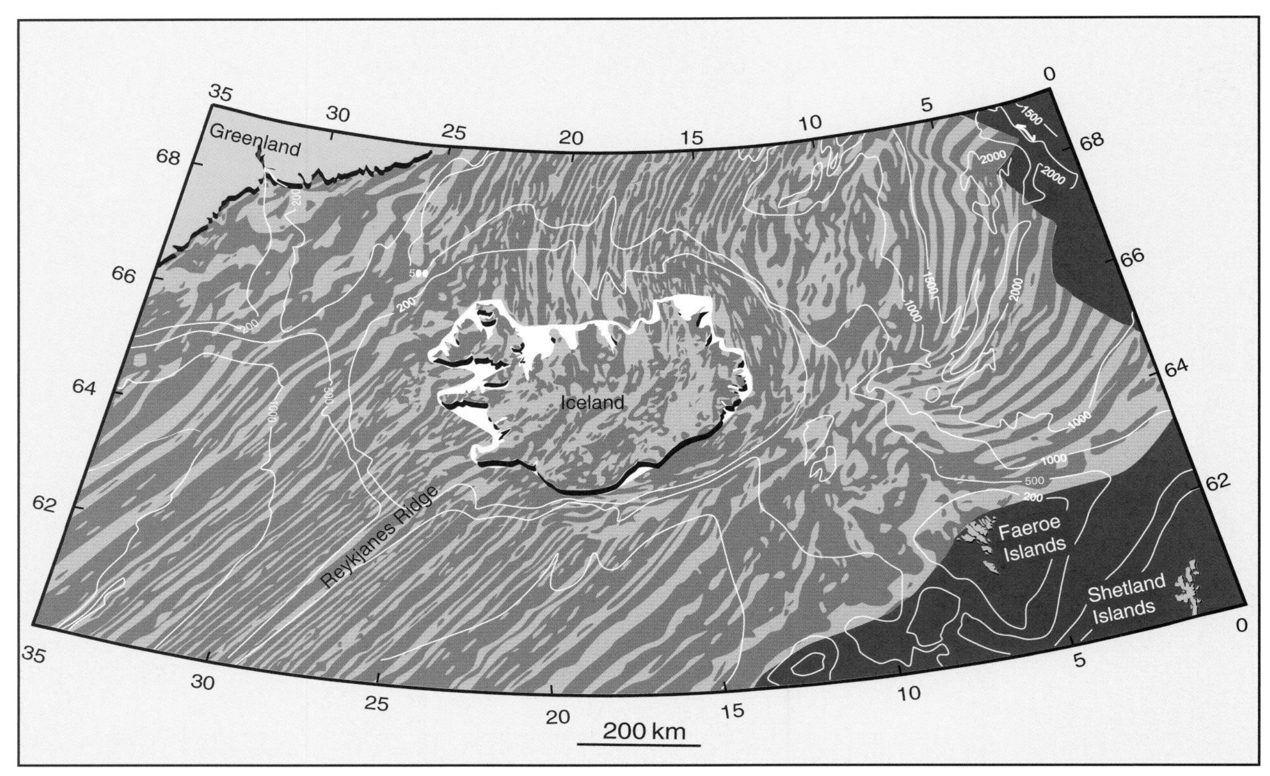

Figure 2.2 Magnetic anomalies are especially well developed in the North Atlantic where the submarine Mid-Atlantic Ridge marks the youngest magnetic zone with normal magnetic polarity (dark gray). Zones with reversed polarity are shown in light gray. Thus a pattern of magnetic stripes with alternating normal and reversed zones is arranged in a symmetrical fashion about the ridge. The youngest zone in the center continues into the Reykjanes Ridge and can be traced across Iceland in the Neo-volcanic zone running diagonally from southwest to northeast. Most of Iceland's active volcanoes are in this zone. After Bott *et al.* (1983).

of earthquakes, known as Benioff Zones, delineated slabs of oceanic crust that are being thrust into the mantle. The volcanic chains are located where these earthquakes are at depths of 150 to 170 kilometers. This situation is seen in all the island arcs, including that of the Mediterranean region (Pichler *et al.*, 1972; Ninkovitch and Hays, 1972).

The southern Aegean volcanic arc extends from the Greek mainland through the islands of Aegina, Methana, Poros, Milos, Santorini, and continues through Kos, Yali, and Nisyros to the Bodrum peninsula of Turkey. It marks the geological boundary between Eurasia and Africa (Fig. 2.3).

At this plate boundary the African plate is descending – or being subducted – into the mantle as it moves northward and converges on the Eurasian plate at a rate of up to five centimeters per year (Figs. 2.3 and 2.4).

How is this related to the volcanism of the southern Aegean arc? The main driving force of plate motions causes relative deformation in the form of extension, compression, adjustment, and shearing in that part of the lithosphere making up the upper 100 kilometers of the earth. Magma can ascend only if extension opens channels through the crust. It can continue to ascend for long periods, intruding at shallow levels where it forms so-called magma chambers that store molten rock in the lower crust. As it resides in these reservoirs, the magma can melt and assimilate crustal material and with time the chemical and physical character of the primary magma is altered. On occasion, this system becomes unstable: the pressure in

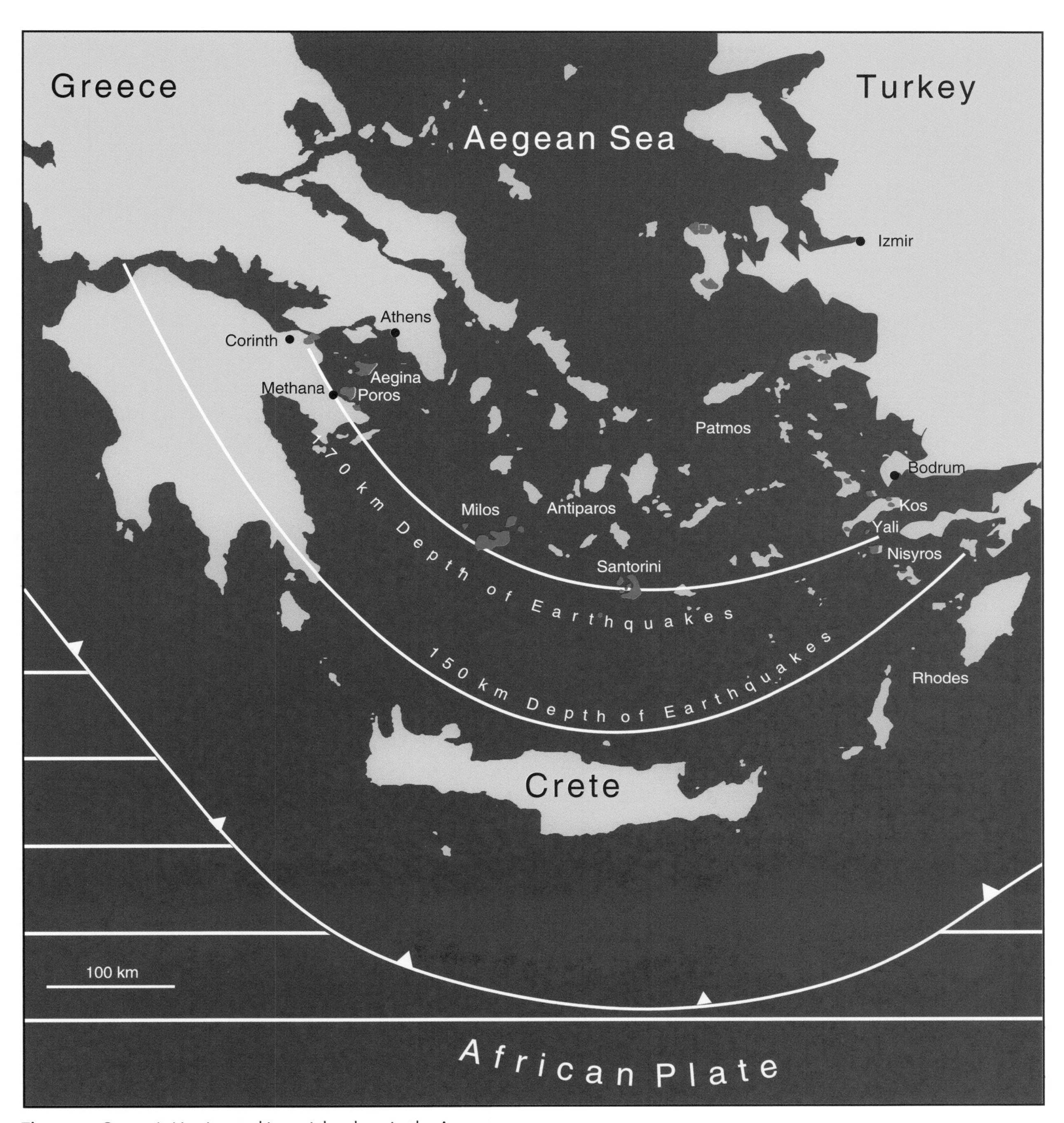

Figure 2.3 Santorini is situated in an island arc in the Aegean Sea, the so-called Hellenic or southern Aegean volcanic arc. The arc consists of a chain of volcanoes that extends from the Greek mainland to the Bodrum peninsula of Turkey. Santorini, the only active volcanic center in this arc, has had three eruptions in the twentieth century alone. The last occurred in 1950. (After McKenzie, 1978.)

Box 2.1

Continental drift

This is a theory that was first seriously proposed by Alfred Wegener, who was the first to lay out the basic principles, in 1915, and who presented a more complete version in 1936. It is based on the assumption that landmasses can drift horizontally on the heavy underlying mantle.

Island arc

This is an arcuate chain of volcanic islands, usually with a deep trench on the oceanward side and a zone of earthquakes that descend from the trench to progressively greater depths beneath the arc. The island arcs of Indonesia, the Antilles, and the Southern Aegean are good examples. Santorini is one of the volcanic islands of the latter.

Plate tectonics

This is a theory first proposed by Arthur Holmes (1944) and Harry Hess (1960) and elaborated by Wilson (1965), McKenzie and Parker (1967), Morgan (1968) and others. It proposes that the lithosphere, consisting of the earth's crust and uppermost mantle, can move on an underlying partially molten zone known as the asthenosphere. The lithosphere is divided into six large plates and a greater number of smaller plates. Oceanic plates are continually being generated by volcanism at oceanic spreading ridges (diverging plate boundaries) and consumed by being thrust back into the mantle where they collide with other plates. On a global scale the rate of generation of new oceanic lithosphere at spreading axes is balanced by the rate of consumption at island arcs and other converging plate boundaries.

Subduction zone

This is a zone where two lithospheric plates collide and one plate is thrust or 'subducted' under the other. The descending plate is usually one consisting of oceanic lithosphere. Earthquakes within and along the upper surface of the subducted plate define a 'Benioff Zone' that descends at an angle of about 45 degrees from a trench to depths of five to seven hundred kilometers in the mantle. Magmas generated near the top of the Benioff Zone rise through the overlying plate to form a chain of volcanoes.

Tsunami

This name is taken from the Japanese term for a sudden, gigantic wave that is triggered by earthquakes and submarine landslides, usually in island arcs and deep-sea trenches. Tsunamis can travel for great distances across the oceans and on reaching a shoreline can devastate coastal areas several kilometers inland. (They are also called seismic waves.)

the magma chamber becomes so great that it eventually exceeds the load pressure and fractures the overlying rocks. Gases dissolved in the magmas begin to escape, much as the gas in a carbonated beverage escapes when the cap is removed, and an explosive eruption ensues.

Five volcanic centers have been recognized in the southern Aegean arc – Sousaki, Methana, Milos, Santorini, and Nisyros. These centers were identified by the geologist von Seebach as early as 1868, who saw that they are associated with a set of cross-fractures. At that time, geologists had no geodynamic concept like plate tectonics to which such observations could be related. We now know that the locations of these centers with respect to fracture systems can be explained in terms of stresses set up by the converging plates. The intersections of fractures are characterized by various types of volcanic manifestations and earthquakes at depths of less than 20 kilometers. The stress patterns can also be seen in the orientations of the five volcanic areas, the morphology of structural features on the seafloor, and the distribu-

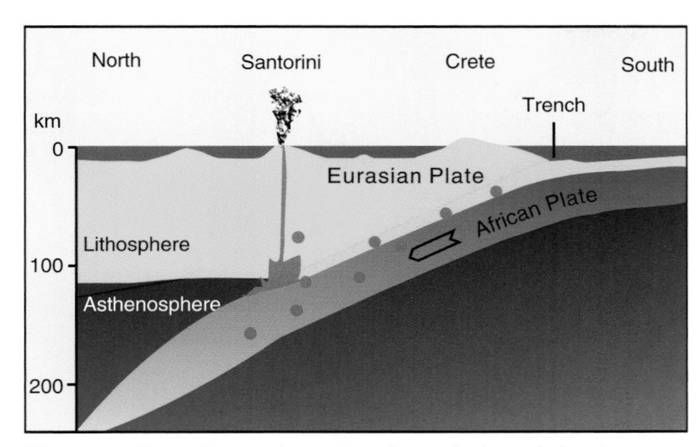

Figure 2.4 A north–south section through the subduction zone near Crete illustrates the African plate descending under the Eurasian Plate. The red dots represent the area where earthquakes are generated at a depth of about 150 to 170 kilometers below the volcanic belt. Modified after Schou Jensen and Håkansson (1990).

tion of earthquakes and the tsunamis they generate (Fig. 2.5) (Papazachos and Panagiotopoulos, 1993).

In the eastern part of the Aegean arc, the volcanic rocks are andesites and rhyodacites of a type that is characteristic of volcanoes on continental margins (Pichler *et al.*, 1972; Keller, 1982). These magmas will be described in greater detail in Chapter 3, but at this point it will suffice to note that they result from the interaction of mantle-derived magmas with the earth's upper crust.

Today, the Aegean region is fragmented into many segments. In a geological sense, it is relatively young (Fig. 2.6). As recently as the Miocene, this region, known as the Aegean landmass, was still structurally coherent, even though faulting beginning around the

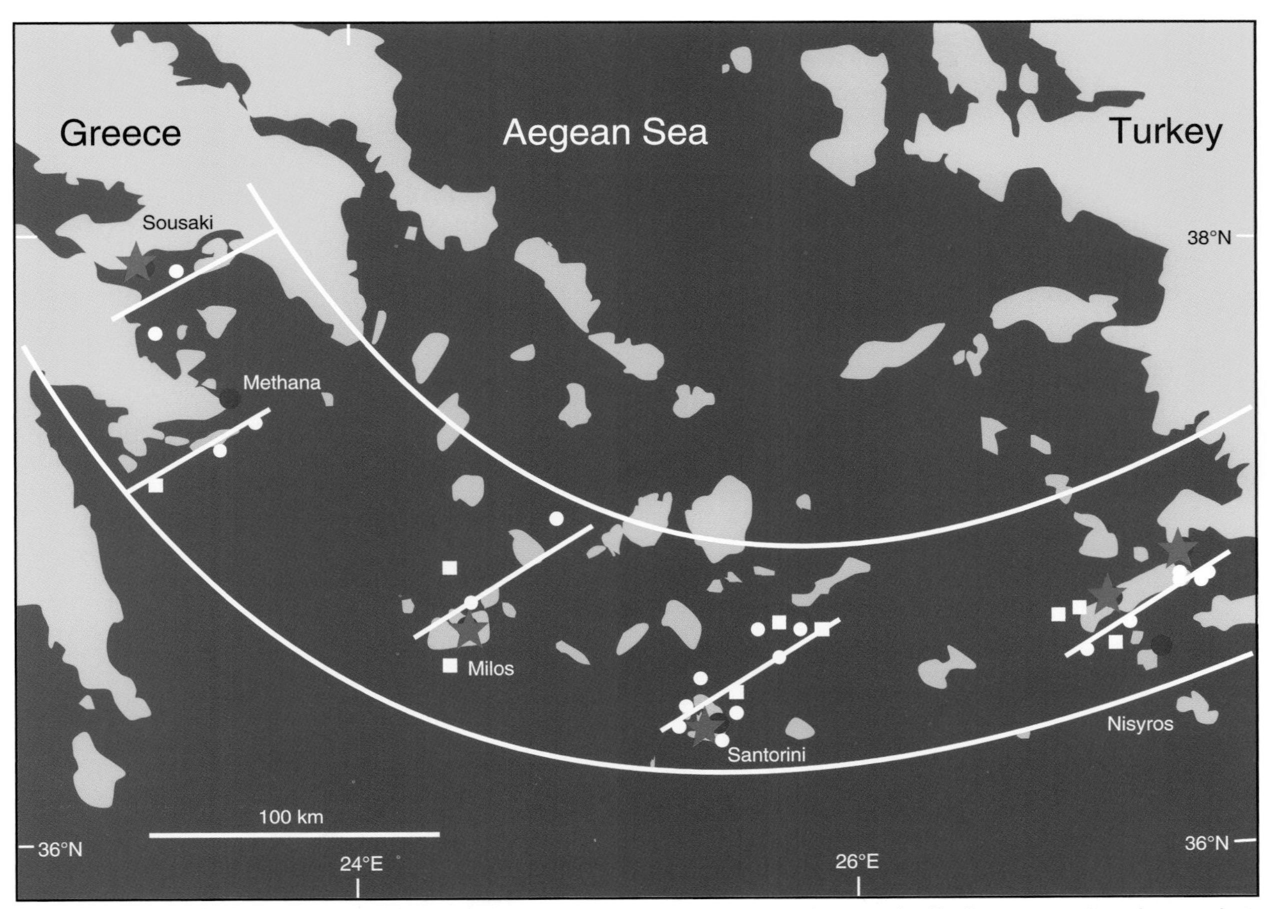

Figure 2.5 Santorini is one of five volcanic centers (stars) making up the southern Aegean volcanic arc. They are connected with tectonic zones of weakness that follow a northeast trend (white lines) and are marked by frequent earthquakes (circles) and solfataras (sulphurous volcanic gases, squares). (After Papazachos and Panagiotopoulos, 1993.)

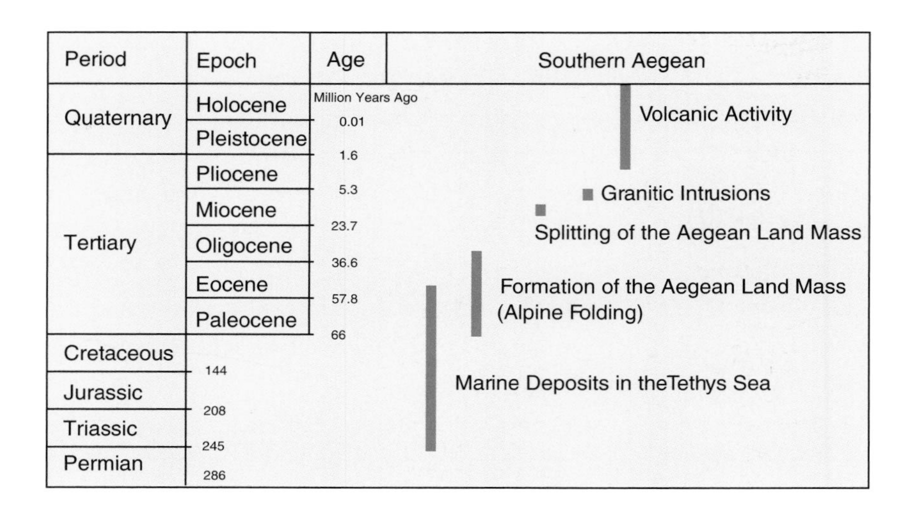

Figure 2.6 Geological events in the southern Aegean Sea from the Permian to the Quaternary.

Box 2.2 Measuring of absolute ages

Most rocks of igneous origin contain small amounts of radioactive elements that can help us to learn when the rock was formed. This is usually done by measuring the proportions of the unstable radioactive isotope and the stable daughter products to which it decays. Knowing the rate of decay, the amount of the daughter product can be used as a geologic clock. In the potassium–argon system, for example, the isotope potassium 40 decays to argon 40. The so-called half-life of a radioactive element is the time it takes for half of the original amount of that element to decay. In the case of potassium 40, this time span is 1.31 billion years.

Table B.2.1. *A comparison of the absolute dating methods*

Method	Isotope	Decay	Half-life (years)	Daughter isotope
Uranium–lead ^{234}U–^{206}Pb	Uranium 238	8 α, 6β	4.50×10^9	^{206}Pb
Uranium–lead ^{235}U–^{207}Pb	Uranium 235	7 a, 4b	7.14×10^8	^{207}Pb
Thorium–lead ^{232}Th–^{208}Pb	Thorium 232	6a, 4b	1.39×10^{10}	^{208}Pb
Potassium–argon K–Ar method	Potassium 40	Capture	1.31×10^9	^{40}Ar
Rubidium–Strontium	Rubidium 87	b	4.72×10^{10}	^{87}Sr
Radiocarbon ^{14}C method	Carbon 14	b	5.73×10^3	Nitrogen 14

end of the Cretaceous (about 60 million years ago) marked the rise of the Alps (Schröder, 1986).

In the Cycladean Massif, e.g. in the metamorphic rocks on Thera, one can see, even today, evidence of structural features that were once very active in this region. Some of these are thrust faults – in which one rock unit was thrust over another – but other types of fault are found as well. The subsidence of separate basins began in the Middle Miocene and was followed by contemporaneous depression of the northern Aegean and the Cretean Sea (around Crete) during Tortonium time (Late Miocene). In the Messinium (also Late Miocene), tectonic movement continued to raise the connection of the Mediterranean to the Atlantic, and as a result, the Mediterranean Sea began to shrink. Evaporite deposits of salt and gypsum were laid down over wide regions. The fragments of gypsum that are occasionally found as xenoliths in the products of the Minoan eruption probably come from these deposits. This so-called salinity crisis lasted only about 500 000 years, and in the Late Miocene the sea connection to the Atlantic was restored. Also during the Miocene, magmatic activity broke out in the Aegean region with the formation of the Cycladean Granite Province which, with renewed movement at the plate boundaries in the Pliocene, led to the intrusion and inundation of the Cycladean Massif (Fig. 2.7). That also set off a new phase of volcanic activity that can be related to the southern Aegean volcanic arc.

During the glacial epochs of the Quaternary this region must have had a totally different appearance than it has today. The Quaternary glaciation led to a lowering of sea levels owing to the huge amounts of water locked up as ice in the polar latitudes. In the Aegean region today only the highest structures rise above the sea as islands, but during glacial times the land surface of the Cycladean Massif was significantly greater. This greatly facilitated the spread of flora and fauna, including people, into the Aegean region.

The volcanic arc is about 500 kilometers long and 20 to 40 kilometers wide. Barberi and his co-workers (1974) suggested an age of 3 million years for the beginning of volcanic activity in the southern Aegean volcanic arc, as was indicated by a potassium–argon age of 3.14 million years obtained for rocks of the Kimolos (Milos volcanic group). This view had been supported by a survey of the ages of the individual centers (Ferrara *et al.*, 1980). Furthermore, Fytikas and Vougioukalakis (in press), noted that the earliest

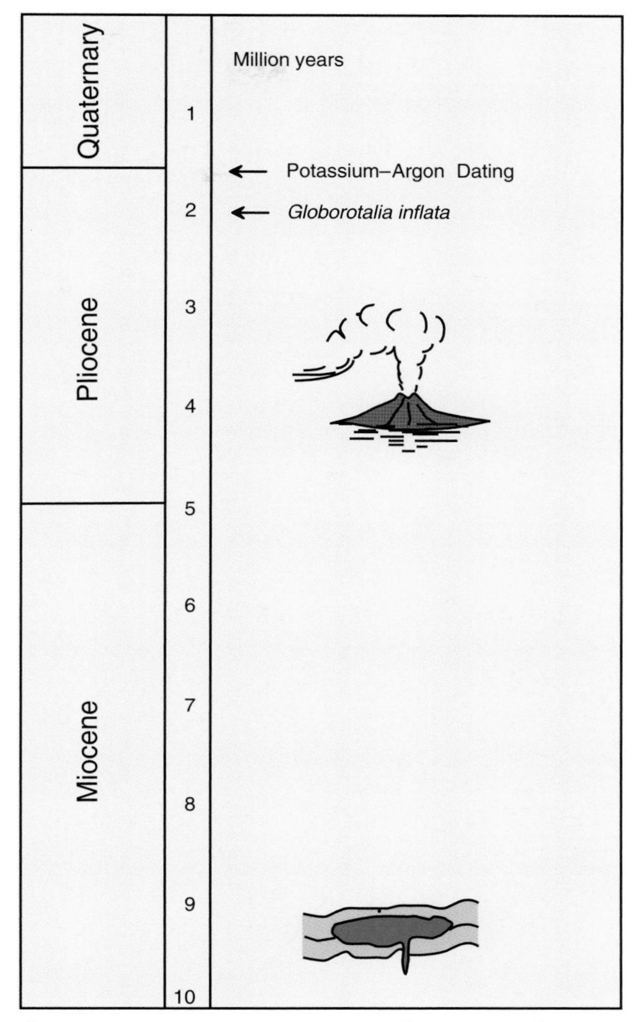

Figure 2.7 Granitic intrusions rose into the metamorphic core of Santorini during the Miocene, but the first known volcanism did not appear until the last part of the Pliocene, when it broke out in shallow water near what is now Akrotiri peninsula. Activity increased during the Quaternary. The time markers indicated are according to potassium–argon dating and the marine fossils used to determine the ages of volcanic rocks and sediments on Akrotiri peninsula.

volcanic structures were formed in the late Pliocene, while the main activity within the southern Aegean volcanic arc took place in the Quaternary. Quite similar results were obtained for Santorini (Seidenkrantz and Friedrich, 1992), where volcanism began in the Late Pliocene (Fig. 2.7). Nevertheless, the greatest amount of volcanic material on Santorini was produced in the last 200 000 years, during which time about twelve volcanic centers have been active.

Volcanism appeared somewhat earlier on the sland of Kos. Keller and his co-workers (1990) reported trachytic ignimbrites of Miocene age (10 to 11 million years old), while dacites and rhyolites of the nearby Kefalos Peninsula have ages of 3.4 to 1.6 million years.

3

The geological development of Santorini

Sedimentary and volcanic deposits of all colors have combined to make Santorini's variegated mosaic. There is a volcanic carapace of lavas, slag, and ash mantles and a core of ancient metamorphosed sediments. In this way, the volcanic island has grown from a succession of very diverse rocks.

Non-volcanic rocks

The highest hill of the island group consists of non-volcanic rocks. It is a relict of the so-called Cycladean Massif, which on Thera includes rocks exposed on Profitis Elias (565 m), Platinamos, Monolithos, Pirgos, and the inner side of the caldera at Athinios, Cape Plaka, and Cape Therma. It is made up of sediments which were deposited in the former Tethys Sea that extended from southern Europe to Sumatra. The deposits were laid down from the Triassic to the Tertiary and were folded, metamorphosed, and uplifted during the Cenozoic Era. Today, these rocks, as well as the sediments derived from them, are now exposed in the mountain chains extending from the Alps through the Himalayas to as far as Indonesia. At Santorini, as on most of the islands of the Cycladean Massif, the original marine sediments have been strongly altered by metamorphism. Limestone has been changed to marble, sandstone to quartzite, and clay stone to mica schists. They have been studied by Fouqué (1879), Phillipson (1896), Neumann van Padang (1936), Papastamatiou (1958), Tataris (1963), Davis and Bastas (1978), Murad and Hubberten (1975), and most recently by Skarpelis and Liati (1990) and Skarpelis *et al.* (1992). The age of the parental rocks has not yet been determined with certainty, but fossils indicate ages ranging from Triassic through Tertiary.

The non-volcanic rocks at Athinios and Cape Therma on the inner side of the caldera are of special geological interest. In these places the original sediments have been metamorphosed to the so-called blue-schist facies. They belong to the Cycladean Blue-schist Belt that also occurs on other islands of the Cyclades. This belt resulted from the intense tectonic deformation caused by collision of two continental plates in Oligocene to Miocene time.

An absolute age of 9.5 million years has been measured for a Late Miocene granite intruded into the metamorphic rocks near Athinios (Skarpelis *et al.*, 1992). The intrusion was also encountered in a bore hole at a depth of 252 meters at Megalochori on Thera (Skarpelis and Liati, 1990). It forms the southernmost rocks in the Cycladean granitic province. The granitic intrusion converted the limestone to skarns with a variety of minerals, including dendritic deposits and veins of pyrophyllite, magnetite, chalcopyrite, and talc. The various colors of these minerals made them a possible source for pigments used in frescoes; they will be discussed more fully in Chapter 10. The lead–zinc deposits, which were mined at the beginning of this century in the inner wall of the caldera at Athinios, were introduced by hot solutions during the closing phases of the granitic intrusion. These minerals also contain silver (Murad and Hubberten, 1975).

The volcanic rocks and their deposition

If one makes a circuit of the caldera by ship and observes the caldera walls of Thera, Therasia, and Aspronisi, one sees a marvelous, multicolored display of horizontally layered rocks with occasional transecting vertical units. At first the imposing picture seems confusing, but if one can interpret the rocks, it is possible to decipher the long volcanic history of Santorini. It is evident from the variety of colors and forms that the volcanic complex was not formed by a single event but by many phases over a long period of time. One notes, first of all, that the volcanic material was laid down as a mantle over the core of ancient metamorphic rocks that, during the Pliocene, stood as an island in the sea.

For about one million years, at least twelve eruptive centers contributed their various products to the volcanic edifice. As it developed, the nature of the rocks changed along with the topographic form of the volcano. As volcanic material was laid down on the land, large parts disappeared into the depths of the sea. At least four times, a caldera was formed only to be filled again with volcanic material. One can trace the various pulses of magma by comparing chemical analyses of the rocks. Druitt *et al.* (1989) reconstructed two major cycles in the rock series and found in them clues to the nature of the magma chamber. These individual eruptive centers and their products are best discussed in order, starting with the oldest and following their development through time (see Fig. 3.1).

In the last century geologists had already observed that the oldest rocks are found on the Akrotiri peninsula (Fig. 3.0), where one can see dacitic lavas that reached the surface through fissures.

In addition, this region underwent large amounts of volcanic uplift that was part of the so-called updoming that connected the land areas to one another. This is especially clear in the case of the Lumaravi–Akrotiri complex and the nearby Archangelos Vouno (Fig. 3.2A). In this region one observes pillow lavas and bright tuffs with still-visible ripple marks. In the 1860s, the geologists Stübel and Fouqué had already noted the felsic marine tuffs of the Archangelos Vouno – the A1 tuff of Pichler and Kussmaul (1980) – in which several occurrences of marine fossils have been used to establish the stratigraphic order of the region (Fig. 3.3). These will be discussed in Chapter 4.

More evidence for submarine volcanism is found at the lighthouse (Fanari) on the southwestern tip of Thera, where one finds red marine conglomerates containing the remains of siliceous sponges similar to our familiar bath sponges (Fig. 3.4). They clearly show us that the activity on the Akrotiri peninsula was the

Figure 3.0 The oldest volcanic tuffs on Santorini record a transition from submarine to subaerial volcanism at Cape Lumaravi. The light tuff of the 1.7 million years old series contains fossiliferous marine rocks that have been domed up and lifted to about sea level by an intrusion of magma. This doming is less than two million years old. The white island of Aspronisi is visible on the right. It is covered with a light layer of Minoan pumice.

Event	Shield formation	Major eruption	Caldera collapse	Magma type	Age (×1000 years ago)
Kameni lava shield				dacite	197BC to 1950(1)
Minoan caldera					
Minoan eruption				rhyodacite andesite	3.6(2)
Riva caldera					
Cape Riva eruption				rhyodacite andesite	18(3)
Therasia lava shield				rhyod., and.	
Eruption of andesites of Oia				andesite	
Upper Scoria 2 eruption				andesite	79±8; 54±3(4)
Skaros lava shield	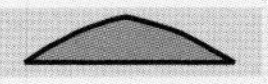			basalt-and.	67±9(4)
Skaros caldera formation					
Upper Scoria 1 eruption				andesite	40 (5)
Vourvoulos eruption				and., dacite	
Megalo Vouno and Kokkino Vouno cinder cones				andesite	76±28; 54±23(4)
Aspronisi tuff ring					
Middle Pumice eruption				dacite, andesite	56±3(5)
Cape Thera eruption				andesite	
Simandiri lava shield				andesite	172±33;172±4(4)
Lower Pumice caldera					
Lower Pumice 2 eruption				rhyodacite andesite	180(4)
Lower Pumice 1 eruption				rhyodacite andesite	203±24(4)
Extrusion of rhyodacites of Cape Alonaki + NE Thera				rhyodacite	257±36; 224±5(4)
Cape Therma eruptions				and., rhyod.	
Extrusion of the andesites of Cape Alai				andesite	586 to144(4)
Cinder cones on Akrotiri peninsula				basalt andesite	626 to 319(4)
Construction of Peristeria complex (domes and lavas)				basalt andesite	536 to 425(4)
Early submarine centers of Akrotiri peninsula				rhyodacite	737 to 563(4), 1600(6)

Figure 3.1 (partly after Druitt 1998).
(1) Historical sources (see Chapter 12). (2) Mean of radiocarbon ages (Friedrich *et al.* 1990). (3) Radiocarbon ages (Friedrich *et al.* 1977, Eriksen *et al.* 1990). (4) Druitt *et al.*, 1998. (5) Radiocarbon ages (AMS method), Knudsen personal communication (see Chapter 4). (6) Paleontological data (Seidenkrantz and Friedrich 1992).

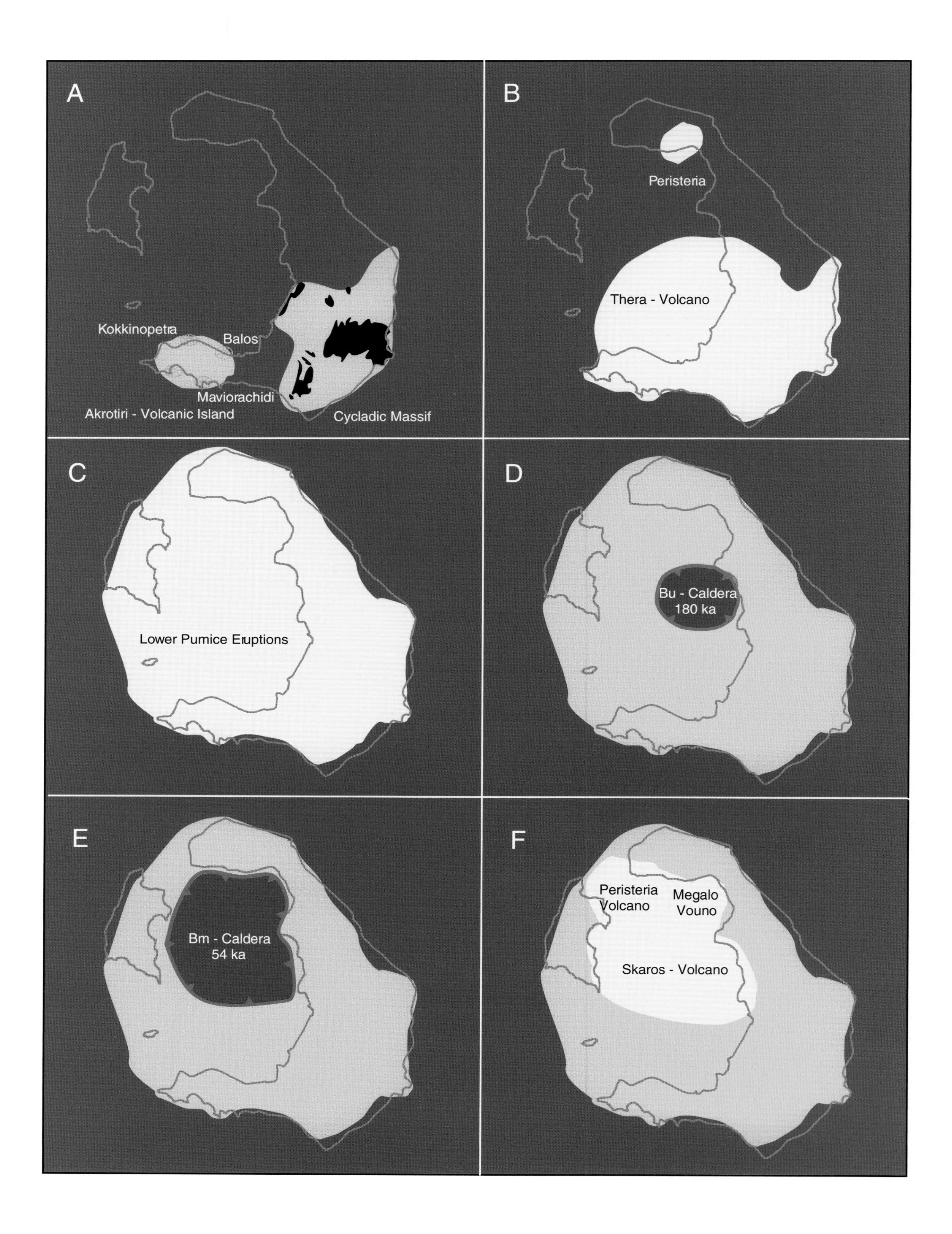
A
Kokkinopetra
Balos
Maviorachidi
Akrotiri - Volcanic Island
Cycladic Massif
B
Peristeria
Thera - Volcano
C
Lower Pumice Eruptions
D
Bu - Caldera
180 ka
E
Bm - Caldera
54 ka
F
Peristeria
Volcano
Megalo
Vouno
Skaros - Volcano

Box 3.1 Volcanic rocks

Basalt

This is a dark fine-grain rock composed mainly of plagioclase, pyroxene, and iron oxides, with or without olivine. It contains about 45 to 53 percent silica. This is the most abundant rock in the earth's crust. It is chemically equivalent to gabbro.

Andesite

This is a gray volcanic rock composed mainly of plagioclase and pyroxene or hornblende. It is named after the Andes where it is the most characteristic product of the large, subduction-related volcanoes. It contains about 53 to 60 percent silica and is chemically equivalent to diorite.

Dacite

This is similar to andesite but contains more silica (60 to 65 percent) and commonly contains quartz. It is named for the Roman province of Dacia (Siebenbergen).

Rhyodacite

This is intermediate between dacite and rhyolite.

Rhyolite

This is a very silica-rich (70 to 77 percent) rock composed of alkali feldspars, quartz, and, in most cases, large amounts of silica-rich glass. It is chemically equivalent to granite.

Ignimbrite

This is a fragmental volcanic rock, normally of dacitic to rhyolitic composition, erupted as molten fragments suspended in hot turbulent gas. It flows across the ground, often covering vast areas. If the flow is still very hot when it comes to rest, the ignimbrite may be welded into a dense, solid rock.

Pumice

This is a very porous volcanic rock, normally of dacitic to rhyolitic composition, consisting almost entirely of colorless siliceous glass and vesicles. It is so light that it will float on water.

Obsidian

This is a very silica-rich volcanic glass, usually jet-black with concoidal fracture surfaces.

Pyroclastic

This is a general term for fragmental volcanic ejecta such as ash and slag.

Tephra

This is a general term for all kinds of pyroclastic material deposited by airfall. Greek word for ash.

Tuff

This is an indurated pyroclastic rock.

Figure 3.2 (Opposite)
A During the Pliocene, Santorini consisted of two islands: a non-volcanic island (the Cycladic Massif) and a volcanic island (Akrotiri). The non-volcanic island was formed by the eroded metamorphic rocks of the Profitis Elias complex, Gavrillos, the area around Athinios, and what is now the small hill of Monolithos. In the Late Pliocene the first eruption broke out southwest of the Cycladic Massif and began to bury it under a mantle of volcanic rocks. The present visible remains of the metamorphic rocks are shown in black. New volcanic areas are shown in light brown. The symbols on Akrotiri volcanic island represent volcanoes.
B In the northern part of Santorini the Peristeria volcano shield began to grow about 500 000 years ago, while Thera volcano connected the non-volcanic islands with the volcanic islands of southern region.
C The two eruptions responsible for the Lower Pumice Series (1 and 2) mantled both the volcanic and non-volcanic parts of Santorini. They occurred around 200 000 to 180 000 years ago.
D During or shortly after the Lower Pumice 2 eruption about 180 000 years ago the roof of the magma chamber under Thera volcano collapsed to form what is referred to as the Bu-caldera.
E Continuous volcanic activity of Thera volcano filled the Bu-caldera. About 60 000 years ago the Middle Pumice (Bm) erupted from a source in the region of the present caldera. Slag and ash of the Upper Scoriae 1 eruption followed this from Megalo Vouno volcano in the northern part of Thera. Around 54 000 years ago the volcanic edifice collapsed a second time, forming the Skaros caldera.
F Lavas of the Skaros shield and volcanic material from the Therasia–Oia shield filled the caldera again. (See Fig. 11.4 for a continuation of this sequence.) Figure partly after Pichler and Kussmaul (1980a, b) and Druitt *et al.* (1998).

Figure 3.3 The updoming on Akrotiri peninsula is especially clear in this view of the plateau and Archangelos Hill. Marine sediments that were formed in shallow water less than two million years ago were uplifted by intruding magma to a height of about 160 meters above sea level.

first volcanism of the Santorini complex. During the Late Pliocene, a volcanic island rose from the sea near the non-volcanic island of Profitis Elias, while submarine volcanic activity produced the lavas and tuffs covering the entire land surface of the Akrotiri peninsula. Here we can see the development of a volcano that started with submarine activity and gradually emerged to erupt on land. This process can also be seen on the nearby Christiana Islands, which are a part of the Santorini complex.

Thus there were volcanoes both above and below the sea, each with their distinctive products. We also know from paleontological evidence that these volcanic events took place around the Plio-Pleistocene boundary at least 1.6 million years ago. The marine tuffs contain cobbles that must be older than the sediments in which they are embedded (Seidenkrantz and Friedrich, 1992). The Akrotiri peninsula was uplifted in the same way as the 'White Island', which rose to the surface near Nea Kameni in 1707. At that time, glowing, viscous magma penetrated sediments on the seafloor and lifted them several hundred meters. The sedimentary cover was broken, and in a few places the lava was extruded through the cracks.

During this same stage, silica-rich, greenish white pumice was erupted together with the early dacitic lavas on Akrotiri peninsula. Recent mapping and new radiometric data provided by Druitt *et al.* 1998 indicate that the submarine centers were formed around 600 000 years ago.

Volcanic activity later shifted northward from Akrotiri peninsula and formed a new shield volcano that was named the Peristeria volcano by Reck (1936). Remnants of this volcano are exposed at Peristeria bay in northern Thera, where the lower two-thirds of Megalo Vouno and the whole of Mikro Profitis Elias are assigned to the Peristeria volcanic complex (Druitt *et al.*, 1998). This region consists of quartz-latitic and

Figure 3.4 On Akrotiri peninsula one still can find remains of the time when Santorini was a submarine volcano and the eruptions still occurred under water. Only a few meters away from the lighthouse (Fanari) one finds red, marine, iron-rich conglomerates containing marine sponges. They are overlain by gray pumice that contains pebbles.

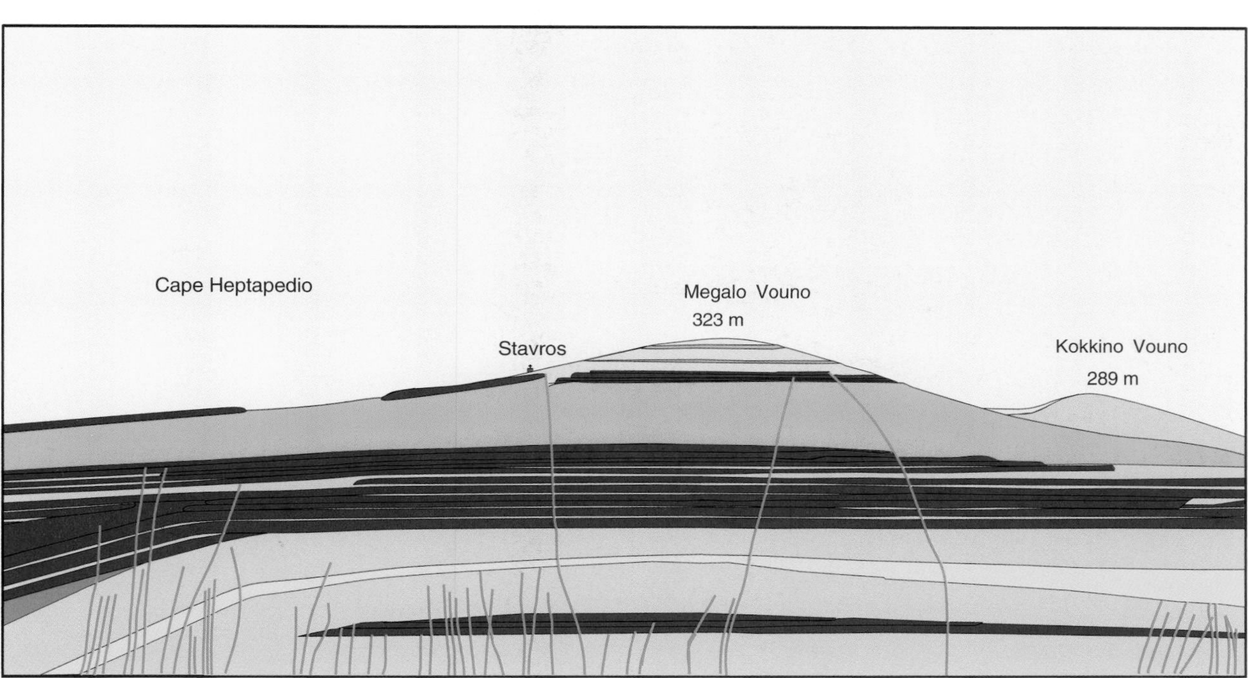

Figure 3.5 On the caldera wall below Megalo Vouno remains of the Peristeria volcano are exposed. They form the lowermost two thirds of the wall, while glassy andesitic lavas build up the upper third of it. In addition, one finds slag and lavas from both Megalo Vouno and Kokkino Vouno. Numerous vertical dykes trending NE–SW transect the wall. They follow a tectonic zone of weakness where magma could arise called the Kolumbo Line (see Fig. 3.9). After Neumann van Padang (1936).

andesitic lavas and volcanic breccias that have been cut by numerous dykes (Figs. 3.5 and 3.6).

Activity continued on Akrotiri peninsula, where at least four small volcanoes produced andesitic and basaltic slag and lavas that were added to the siliceous products of the earlier eruptions. The sources for the more silica-poor, mafic material were the volcanoes of Balos, Kokkinopetra, Mavro, and Mavrorachidi. The volcanoes Balos, Kokkinopetra ('Red stone'), and

Figure 3.6 A single volcanic dyke below Megalo Vouno transects the caldera wall. Some of these so-called feeder dykes expand in the upper part of the caldera wall and form a horizontal layer of lava.

Mavrorachidi consist of basaltic andesite and were recently dated by Druitt *et al.* (1998) using the K–Ar methods (Fig. 3.1).

The lavas and tephra of the Thera Volcanic Complex formed the next major unit. Its center was in the present caldera in the region of the present Kameni Islands. The material that erupted from this center made up the dominant part of the succession in the following period of volcanic activity. It covered the small islands and joined them into a single, large volcanic complex. These rocks are easily seen today in the caldera wall from Cape Fanari to Cape Turlos and in the caldera walls of Thera. Lavas are found in only a few small exposures in this series and are limited to the region from the harbor of Fira to some hundred meters south of Cape Alonaki.

The Cape Therma Series forms a thick sequence of ejecta which is especially well displayed in the southern part of Thera. The thickest part is the Cape Therma-1 sequence, which rests unconformably on the metamorphic basement. Seward *et al.* (1980) determined the age of this series (about 0.9 to 2 million years) using fission track methods. The pyroclastic unit can be correlated with the volcanism of Akrotiri. In the Cape Therma-1 series one finds pumice with a dacitic composition. At Athinios, it reaches a thickness of 2.4 meters. The Cape Therma-2 eruption produced rhyodacitic pumice falls, which can be seen in discontinuous exposures between Cape Katofira and Akrotiri peninsula. The Cape Alonaki rhyodacitic lavas with an age of 257 ± 36 thousand years (Druitt *et al.* 1998), and the Cape Therma-3 deposits overlie this unit. On the latter, there lies a red-black ignimbrite with dioritic and gabbroic inclusions.

The Lower Pumice (Bu) is the next major unit that is especially conspicuous in the caldera wall on Thera. It is a white pumice about 30 meters thick and extending from Cape Akrotiri to Cape Turlos; it is also found in thinner layers in the northern part of Thera (Fig. 3.2C). Neumann van Padang (1936) designated it the 'Unterer Bimsstein', abbreviated Bu. The Bu series consists of two pumice layers, namely Bu_1 and Bu_2. Both Bu eruptions started with pumice falls and ended with an ignimbrite (ash flow), and both units are classified petrographically as rhyodacite. It has been possible to locate the eruptive sources of both units from the variations of their thickness and grain size. The source of Bu_1 was in the southern part of the present island of Thera (Fig. 3.7), whereas the source of Bu_2 was more to the north in the area of today's Kameni Islands.

The lavas of Skaros, Mikro Profitis Elias, and Megalo Vouno volcanoes unconformably cut the Bu series. The Bu series can be seen again in the lower part of the caldera wall below the town of Oia. The disruption of the series at these places is obviously related to the formation of the Bu caldera (Pichler and Kussmaul, 1972). This caldera was formed by volcano-tectonic collapse following evacuation of the magma reservoir in the Bu_2 eruption. This ancestral caldera was later filled with the products of a succession of later eruptions. The lower pumice layers of Santorini were dated at around 100 000 years (Seward *et al.*, 1980). New Ar 39–40 ages measured by Druitt *et al.* (1998) give an age of 203 ± 24 thousand years for the Bu_1 layer, while the Bu_2 layer is dated at about 180 000 years.

The two Bu ashes are spread over a wide extent of the Aegean region where they form useful time-marker horizons. Judging from the visible thickness of the Bu ash and its wide distribution in drill cores in the Aegean Sea, one can conclude that the Bu eruptions were much stronger than the Minoan eruption.

Simandiri formed after an interval of repose of unknown duration when new activity began within the Bu caldera. The construction of the andesitic Simandiri lava shield, referred to by earlier workers as the Megalo Vouno Complex, has recently been dated at 172 ± 33 thousand years (Druitt *et al.*, 1998). It con-

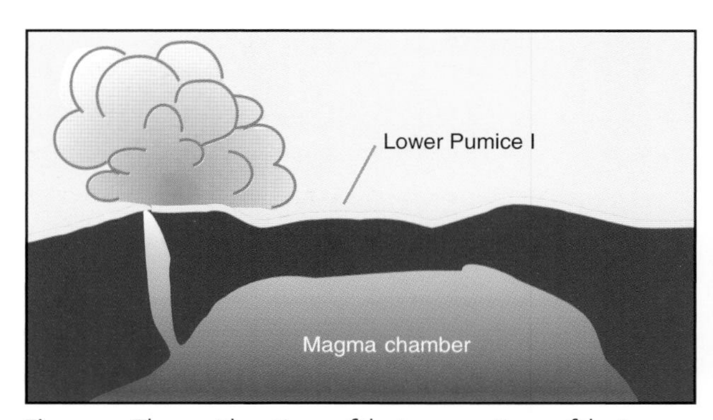

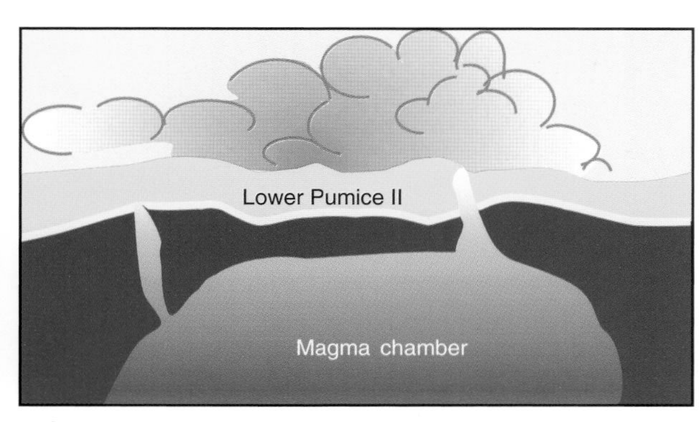

Figure 3.7 The vent locations of the two eruptions of the Lower Pumice (Bu_1 and Bu_2). (After Günther and Pichler, 1973.)

sists mainly of andesitic and latite–andesitic lavas and slag. The shield, which extended mainly through the northern parts of Thera and Therasia, consists of a variety of lavas and pyroclastic beds. The so-called northern lavas and ash came from several volcanic centers: Megalo Vouno, Skaros, and the Simandiri lava shield. The products of these volcanoes were disrupted when the Cape Riva and Minoan calderas were formed.

The Middle Pumice layer was formed by a strong eruption from Thera Volcano. It is seen below the town of Fira on both sides of the harbor. One finds

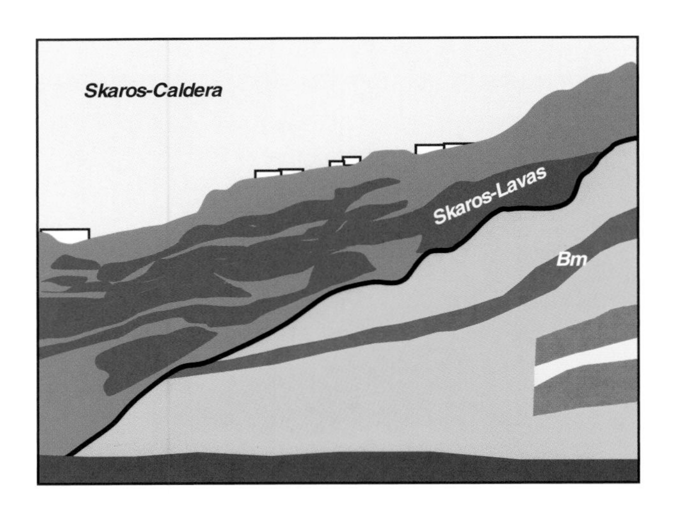

Figure 3.8 The lavas of the Skaros caldera (top) that were formed about 60 000 years ago cut the older rocks of the Middle Pumice (Bm). The line of discordance runs diagonally through the photo, which was taken beneath the town of Merovigli on Thera.

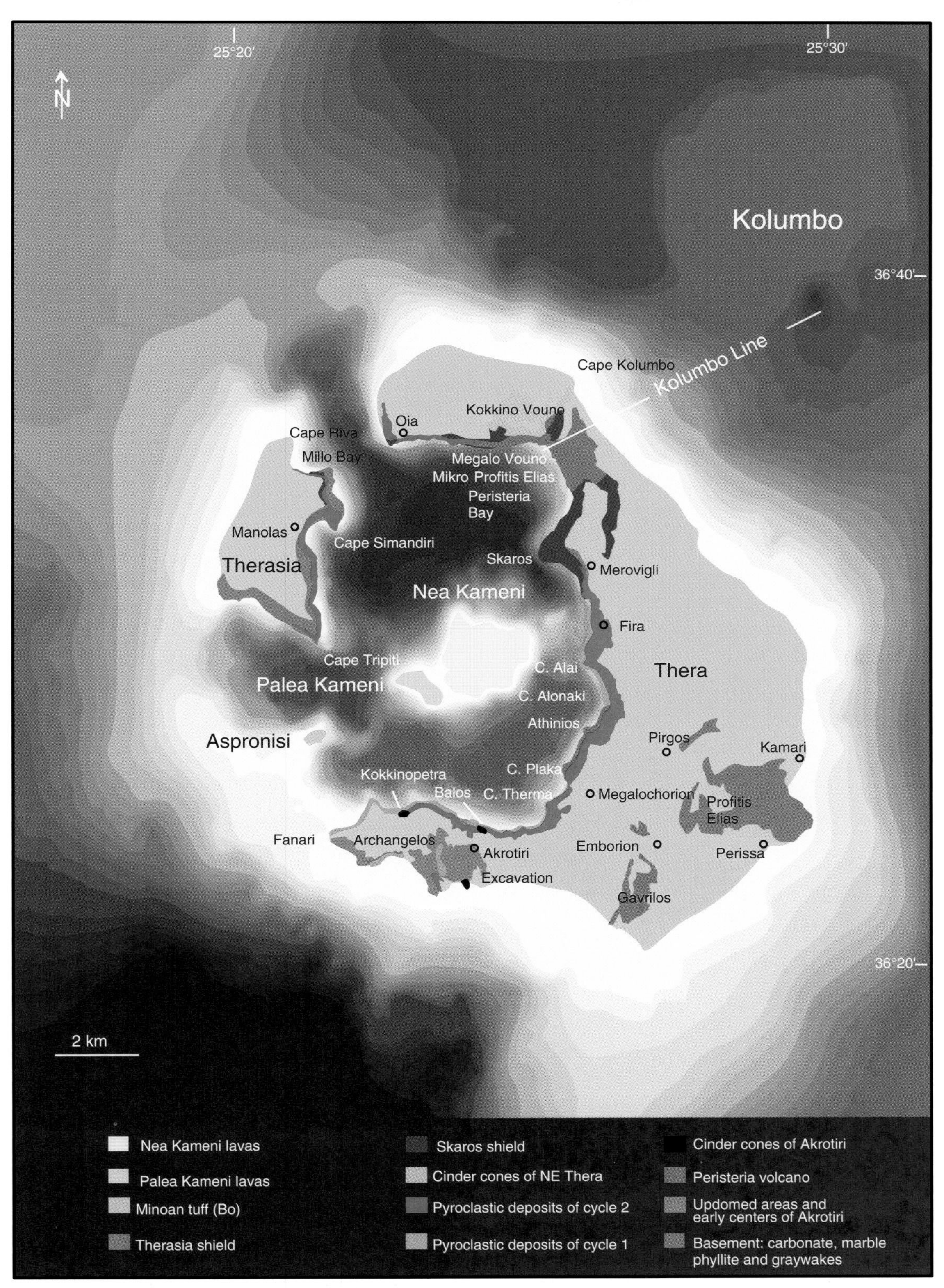

Figure 3.9 A simplified geological map showing also the bathymetry and locality names and towns. (After Druitt *et al.*, 1998.)

there a black glassy dacitic horizon that has been displaced by a fault near the steps leading from the harbor to Fira. Günther (1972) named this layer 'Middle Pumice' (Bm for Mittlerer Bimsstein).

The Bm layer can easily be traced for several kilometers in exposures in the caldera wall of Thera. Below the village of Merovigli, the layer is abruptly cut off by the lavas of Skaros (Fig 3.8). It forms a prominent pyroclastic marker horizon but has a strong lateral variation. In the caldera wall below Merovigli, the Bm layer is a strongly welded, massive red air-fall tuff but below Fira it is pitch black. It gradually changes color and structure within a few kilometers southward from brownish welded pumice to bright pumice. From this, it has been concluded that the eruption center of the Bm layer must have been within the present caldera in the vicinity of Merovigli. The age of this unit has now been dated at 56 050 ± 2980 years BP using AMS radiocarbon method (Knudsen, personal communication).

Figure 3.10 Erosional channels in the quarry south of Fira on Thera are filled with tuff and slag of the Upper Scoria 2 eruption. They transect the uppermost tuff layers on Thera and slope in the same direction as the volcanic edifice.

Stratigraphically, the Middle Pumice is followed by a tuff ring on Aspronisi and an eruption of the cinder cones of Megalo Vouno and Kokkino Vouno; the latter have been dated at 76 ± 28 and 54 ± 23 thousand years (Druitt *et al.*, 1998). The andesitic lavas of the Vourvoulus eruption and the Upper Scoria 1 eruption overlie the cinder cones. Upper Scoria 1 contains ashes with plant remains. Charcoal from two concordant layers gave ages of 54 250 ± 700 BP, and a very distinctive wavy layer, called the 'double band', has an age of 40 500 ± 70 BP. The double band contains plant remains that are described in Chapter 5.

This long period of growth was followed by a destructional phase when the Skaros caldera was formed. This caldera, which was similar to the earlier one formed after the Bu eruptions, was subsequently filled by products of the Skaros shield volcano. A series of ashes belonging to the Upper Scoria 2 eruption contains plant remains, which were dated using radiocarbon methods at about 40 000 years. Ages of 79 ± 8 and 54 ± 3 thousand years were obtained using K–Ar ages and Ar 39–40 methods respectively. These conflicting data must be taken with a grain of salt, since these dating methods are near the limit of their useful range. Of the two, however, the radiocarbon age of about 40 000 years seems more reliable, because it is based on several measurements and different materials. According to Druitt *et al.* (1998), some of the andesites of Oia overlie the Upper Scoria 2 unit stratigraphically and are followed by the Therasia lava shield.

The Skaros volcano consists, in its lower part, of andesitic lavas, while its upper part is made up of a series of thinner andesitic lava flows (Figs. 3.7 and 3.8). Dacitic lavas and domes covered them. The lavas of Skaros abut abruptly against those of Mikro Profitis Elias.

Glassy andesitic lavas formed the upper third of Megalo Vouno. In addition, one finds there slag and lavas from both Megalo Vouno and Kokkino Vouno. The stratigraphic relationship found here shows that these lavas are older than the Skaros shield complex.

During this same period of activity at Megalo Vouno, a maar (a vent) was formed at Cape Kolumbo (Pichler and Kussmaul, 1980).

The upper ignimbrite, also known as the Cape Riva ignimbrite, forms a marker about seven meters thick. It has been correlated with the U-2 tephra of marine borings (Keller, 1981). The age of this ignimbrite is estimated at about 21 000 years BC on the basis of radiocarbon dating of charcoal (Friedrich and Pichler, 1976; Friedrich *et al.*, 1977; Eriksen *et al.*, 1990). This age is considered to be quite good, for it is based on charcoal remains of several small trees which were buried by the ignimbrite. (see Chapter 5, Fig. 5.0). The distribution of this ignimbrite has been described by Pichler and Kussmaul (1980) and by Druitt (1985). The original C^{14} dating gave ages of 18 000 years, but as a result of the new calculation curve of Bard *et al.* (1990), it is now taken as 21 000 calendar years BP.

After deposition of the Cape Riva ignimbrite, the Skaros shield collapsed and formed the Riva caldera.

Figure 3.11 A lateral section through the upper part of the Thera volcano in the quarry south of Fira. In the uppermost part of the photo, the light pumice of the Minoan eruption can be seen filling an old erosional channel, while the lower half shows the deeply weathered deposits of the Upper Scoria Series. In the lower third one sees the characteristic, wavy 'double-band' that has a radiocarbon age of about 40 000 years. It contains plant fossils and is an air-fall deposit. The light first layer of the Minoan eruption (Bo_1) mantles the bottom of an old erosional channel. It has a uniform thickness as is typical for air-fall deposits. This is in contrast to the two following layers of this eruption (Bo_2) and Bo_3). The latter lie horizontally, thus proving that they were formed by base surges (Bo_2) and ash flows (Bo_3) moving laterally.

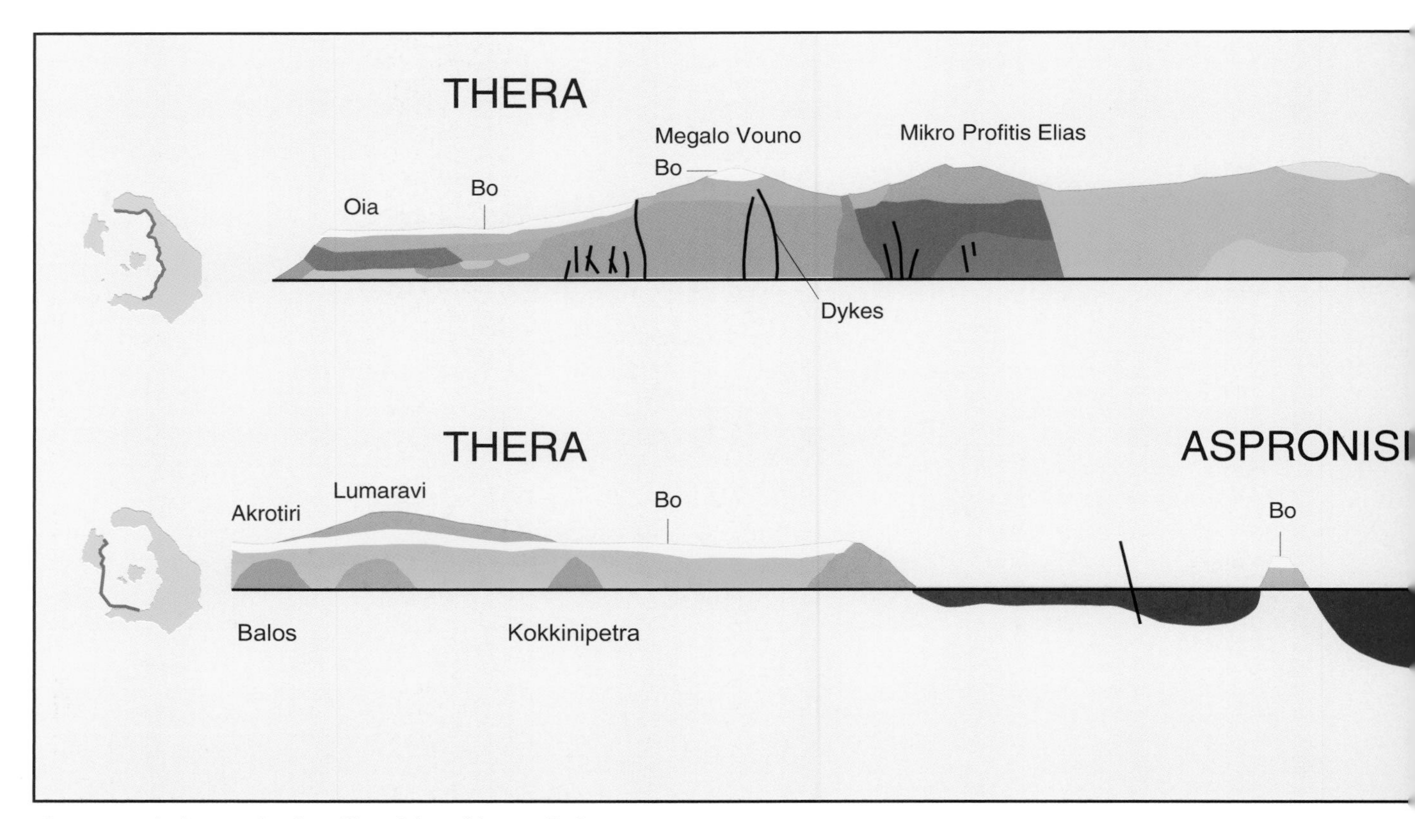

Figure 3.12 The longitudinal profiles of the caldera wall of Santorini show the development of the volcanic edifice with its constructional and destructional phases. Modified after Neumann van Padang (1936).

This caldera has also been called Stronghyle caldera by Eriksen *et al.* 1990.

The Minoan ash layer is the uppermost layer on Thera, Therasia, and Aspronisi. It is also called the Upper Pumice or Oberer Bimsstein (Bo). This layer is an important time horizon for both archaeologists and geologists, since it covers the Bronze Age settlements and forms a thick ash layer over much of the southeastern Aegean Sea (Figs. 3.9 to 3.12). It will be discussed in more detail in Chapter 6. During the last phase of the Minoan eruption the volcanic edifice collapsed again to form the Minoan caldera. Then, starting in 197 BC, the caldera-filling process began again and with time produced the Kameni shield with the islands of Palea and Nea Kameni.

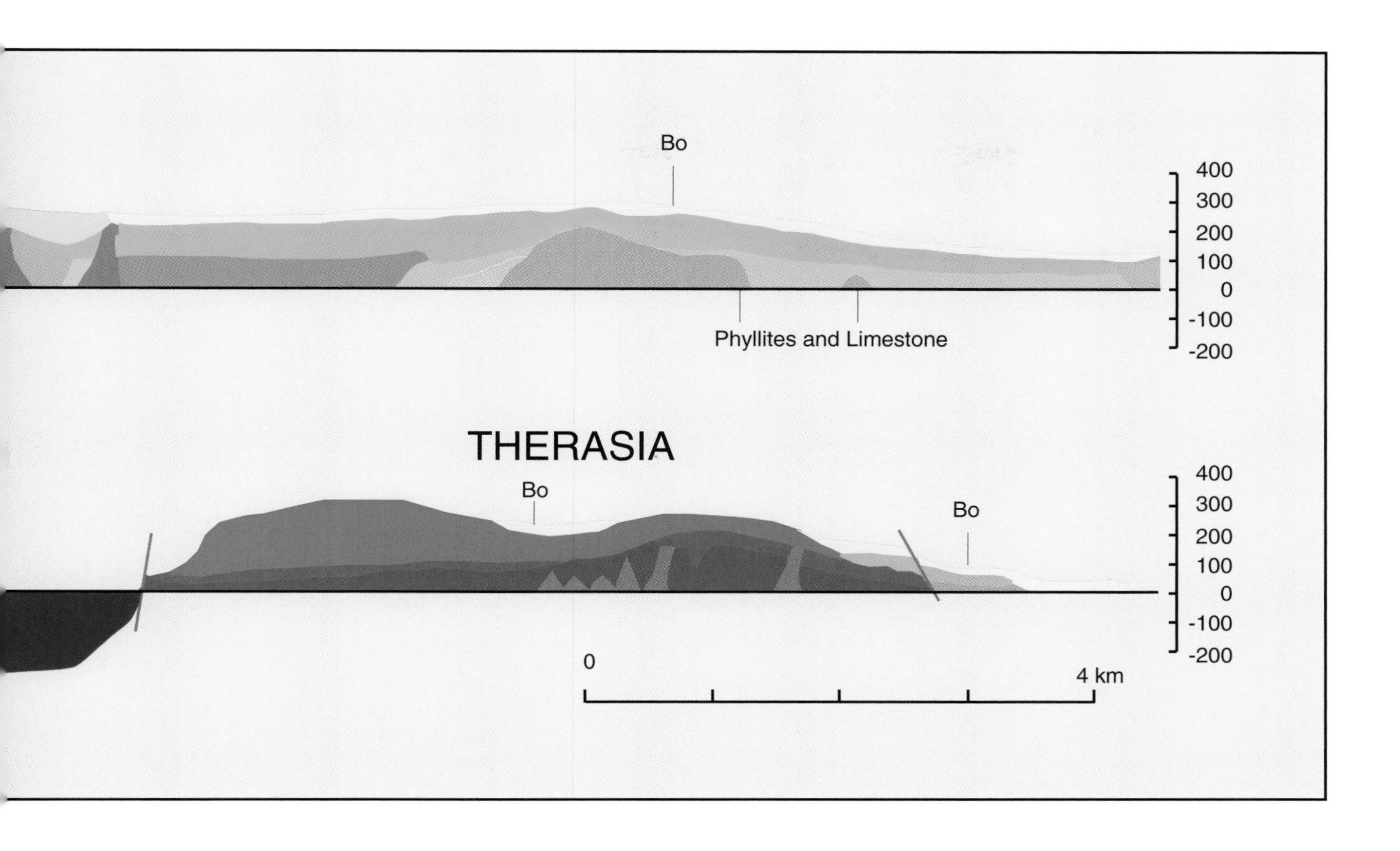
Bo
400
300
200
100
0
-100
-200
Phyllites and Limestone
THERASIA
Bo
Bo
400
300
200
100
0
-100
-200
0
4 km

4

Stratigraphy – the backbone of geology

The layers of volcanic ash erupted from Santorini have a wide regional importance, because they were produced by a relatively brief but well-defined event and were laid down contemporaneously in a wide region. Knowing their ages, one can use them as time markers wherever they can be identified. Determining the ages of stratigraphic units of this kind is one of the important challenges facing the

Figure 4.0 The caldera wall between Fira and Athinios contains a geological record reaching back about 200 000 years. In the lower levels can be seen the double layers of the Lower Pumice (Bu); near the top of the dark section the Middle Pumice (Bm) is visible. This is followed by the plant-bearing double band, which is too thin to be seen in this view, and at the top are the three layers of the Minoan eruption (Bo).

Figure 4.1 (right) A schematic diagram, which shows the most important dated layers on Santorini, ranging in age from Triassic to Quaternary. From Seidenkrantz and Friedrich (1992).

geologist. To do this, the dating tools of stratigraphy and paleontology, fossils and the basic laws of physics, are used.

Paleontological remains

The geologist seeking to determine the age of sedimentary rocks of a certain region usually relies on certain key fossils that establish the stratigraphic sequence. They can provide a valuable tool that is usually quicker and less expensive than absolute dating. Unfortunately, however, fossils are of little use in studying the oldest beds of Santorini, because they are rare and poorly preserved. Consequently, the stratigraphic relations of the non-volcanic parts of the islands are not very well defined.

Marine fossils of the Profitis Elias Massif

The metamorphic rocks exposed on Profitis Elias on Thera are exceptionally fossil poor. Despite the wealth of fresh exposures made over the last decade, only a few impressions of mussels (Megalodontitdae) have been found (Papastamatiou, 1958). They make it possible, however, to place the beds in a general age succession within the time period of the Late Triassic (Fig. 4.1). Mussels of this kind are also known from the nearby islands of Naxos and Mykonos.

An additional fossil discovery, also in the Profitis Elias Complex, shows that even younger marine beds occur in the region. Tataris (1963) described fossils in a lens of limestone within weakly metamorphosed

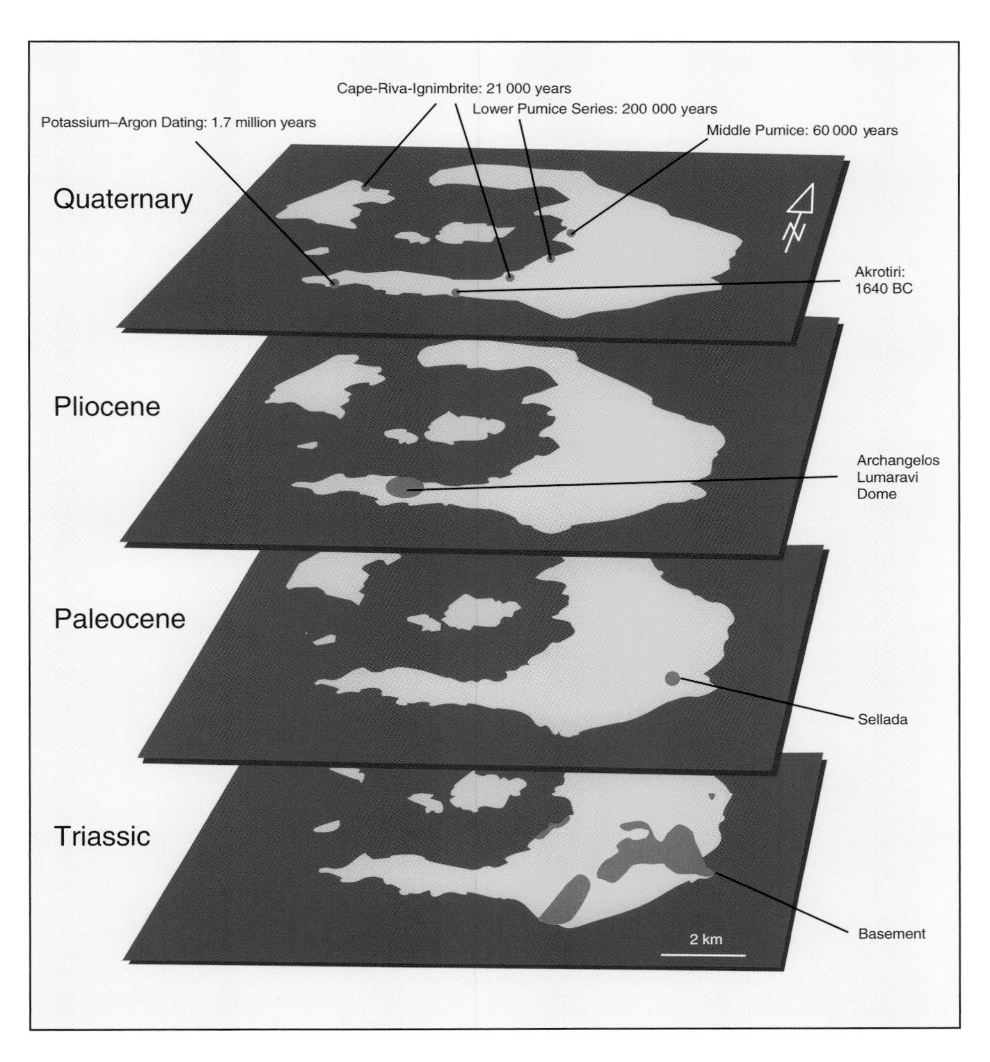

conglomerates at Sellada in the saddle between Profitis Elias and Mesa Vouno. He found foraminifera (Miliolidae, *Laffiteina* sp.) as well as snails and other fossils that can not be identified with certainty. The faunal assemblage indicates that these beds belong to the Cretaceous and probably the Paleocene (Montium). According to Tataris, there are similar sediments on the nearby island of Anaphe that contain fossils, which make an Eocene age more likely.

Marine fossils at Archangelos Vouno

On Akrotiri Peninsula in the southern part of Thera marine beds underlie products of the earliest known volcanic phase of Santorini. They have been raised some 200 meters above sea level by the updoming effect of intruding magma. They are especially well exposed at Archangelos Vouno, Lumaravi Vouno, and in the caldera wall at Cape Lumaravi (Fig. 4.2). They contain fossil remains, listed in Box 4.1, that have been identified as Late Pliocene.

Neumann van Padang collected fossils from this same locality at Archangelos Vouno. They were identified by Quenstedt (1936), who reported that *Pecten jacobaeus* Linné and *Pecten septemradiatus* Müller were among them. From the northern side of Lumaravi he mentioned *Pecten varians* and *Pinna* sp. (probably *Pinna pectinata* Linné).

A new collection that I was able to make from the same place on Archangelos Vouno yielded only very poorly preserved remains of mussels (*Pecten*) some of which had molds of foraminifera. The identification of these fossils was made by Marit-Solveig Seidenkrantz of Aarhus, who had also made a collection (Seidenkrantz, 1989).

The new studies of the foraminifera at Archangelos Vouno show that the fossil-bearing sediments were laid down in a near-shore setting. This interpretation is supported by the occurrence in the sediment of coastal debris derived from volcanic rocks. The foraminifera are shallow-water forms, which were originally laid down in water with a depth of up to 25 meters (Fig. 4.2).

Another occurrence of fossils on the Akrotiri peninsula is located at Cape Lumaravi. It contains a fauna of foraminifera that lived in relatively deep water (more than 100 meters). One finds mostly planktonic forms, i.e. forms that drifted in the water (Seidenkrantz and

Box 4.1 Macrofossils of Archangelos Vouno on Akrotiri Peninsula (from von Fritsch, 1871, page 176)

Schizaster minor Mayer
Terebratulata vitrea L. sp.
Ter. septata Phil. valv. sup. (Waldheimia)
Ter. euthyra Phil. (Waldheimia)
Ter. caput serpentis L. (Terebratulina)
Ostrea hippopus Lam.
Anomia patelliformis E. sp. (Placunanomia)
Pecten similis Laskey
Pecten septemradiatus Müll. (= *pseudamusium* Chemn.)
Pecten varius Penn.
Avicula sp.
Arca barbata L.
Arca pectunculiformis Scac.
Nucula sulcata
? *Leda nitida* Brocchi sp.
Cardium edule L.
Cardium roseum Lam
Lucina Astensis Bronn
Lucina spinifera Montf.
Venus, Cytherea or *Circe,* sp. nov.
? *Venus*
Venus gallina L.
? *Cardita*
? *Corbula*
Dentalium tetragonum Brocchi
Dentalium Dani Hark
Turbo sanguineus L.
? *Rissoa*
Assiminea littorina Delle-Chiaje
Vermetus glomeratus L.

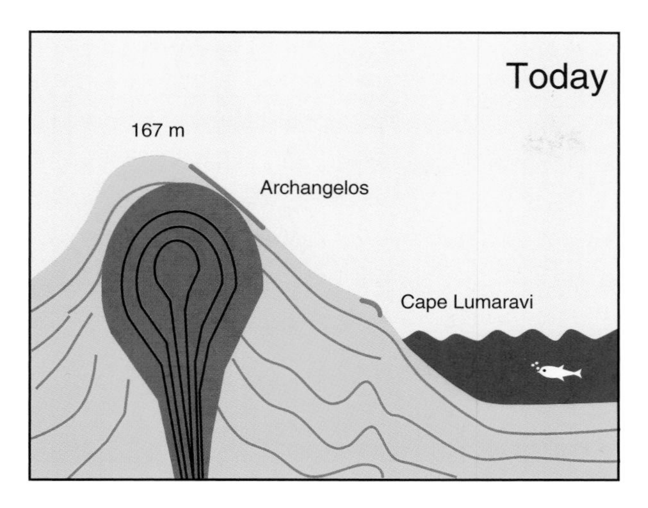

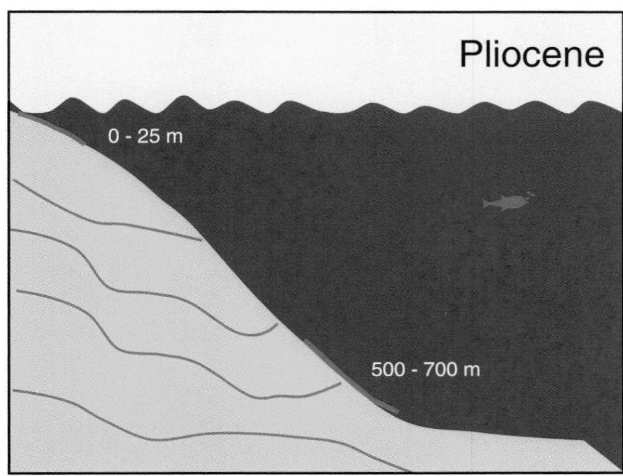

Figure 4.2 The lower part of the diagram shows the original water depth in the Lumaravi–Archangelos area during the Upper Pliocene. The thick red lines mark levels of marine biotopes. The upper diagram shows the doming and uplift of the marine layers that occurred at the end of the Tertiary period.

Friedrich, 1992). This fauna is at most two million years old. The minimum age has been given by absolute dating as 1.6 million years. So the fossils date from around the Pliocene–Pleistocene boundary. Since erosional debris of volcanic origin can also be found in the fossil-bearing beds at Archangelos Vouno, one can conclude that the uplift of the marine beds had already occurred before the deposition of the fossils, probably also in Upper Pliocene. This relative dating gives us a crude indication of when volcanism began on Santorini.

Absolute dating

As one can deduce from the fossils named above, the first volcanic event at Santorini was probably in the Upper Pliocene. This means that one can use the potassium–argon method for measuring ages, provided, of course, that the rocks to be dated are fresh and contain adequate amounts of the radioactive isotope of potassium.

When this method of dating was used for lava on another island of the South Aegean arc, Kimolos (Milos Group), it gave an age of approximately 3.5 million years. When used for rocks on the island of Thera, a thick dacitic lava flow at Cape Akrotiri and a submarine breccia at Cape Mavro gave an age of 0.3 to 0.6 million years (Druitt *et al.* 1998).

The fission-track method, described in Box 4.2, has been used successfully to determine the age of crystals of the mineral zircon found in the lower pumice layer. It gave an age of one million years. When fragments of obsidian (dark volcanic glass) that occur as exotic inclusions in the same lower pumice were dated they indicated an age of about 100 000 years (Seward *et al.*, 1980). The lower layer of pumice (Bu), which has a thickness of about 40 meters on Santorini, is also found in many cores taken from

Box 4.2 The fission-track method

With this method it is possible to determine the age of certain commonly occurring minerals, such as apatite and zircon. It is based on the principle that from the time crystals are formed they are subject to radiogenic emissions resulting from the steady decay of trace amounts of radioactive elements that are contained in the crystals. The emissions produce small tracks that can be made visible in a thin section of a rock by etching them with acid. By counting these tracks under a microscope the age of the minerals is obtained using a simple relationship that defines the rate at which tracks are produced with time: the more tracks the older the mineral.

sediments on the floor of the Aegean. This makes it a useful time horizon and illustrates the importance of Santorini in the stratigraphy of the Mediterranean. In the past few years new dating of this kind, especially that of Druitt *et al.* (1998), has led to a number of new interpretations of the regional stratigraphy. They dated the two units of Lower Pumice series to 200 000 and 189 000 years respectively.

Dating of organic remains

The first subaerial lavas probably appeared in Upper Pliocene. The pollen that Sauvage and Jarrige (1978) were able to identify in sediments on the Akrotiri peninsula are probably of the same age. Organic remains of this kind cannot be dated by the radiocarbon methods, because this technique is reliable only for ages of up to 50 000 years. But the flora flourishing in the Quaternary left numerous datable remains in the ash beds, and these provide important stratigraphic and climatic information.

In volcanic regions, such as Hawaii or Iceland, one is always surprised to see that shrubs, trees, and grass leave vestiges, especially in the form of impressions or even organic remains. These remains are

Figure 4.3 The Cape Riva ignimbrite occurs in many places on Santorini. It is a very important marker horizon, since it is precisely dated and is found as tephra outside Santorini. The Theran people of the Bronze Age used this ignimbrite for the walls of their houses, as seen in the excavations at Akrotiri and Potamos. In the upper picture one can see the ignimbrite dipping gently toward the caldera just above the seated person. The lower picture shows a characteristic feature of the ignimbrite, its so-called fiamme structures. Both photographs were taken in the Mavromatis quarry near Akrotiri.

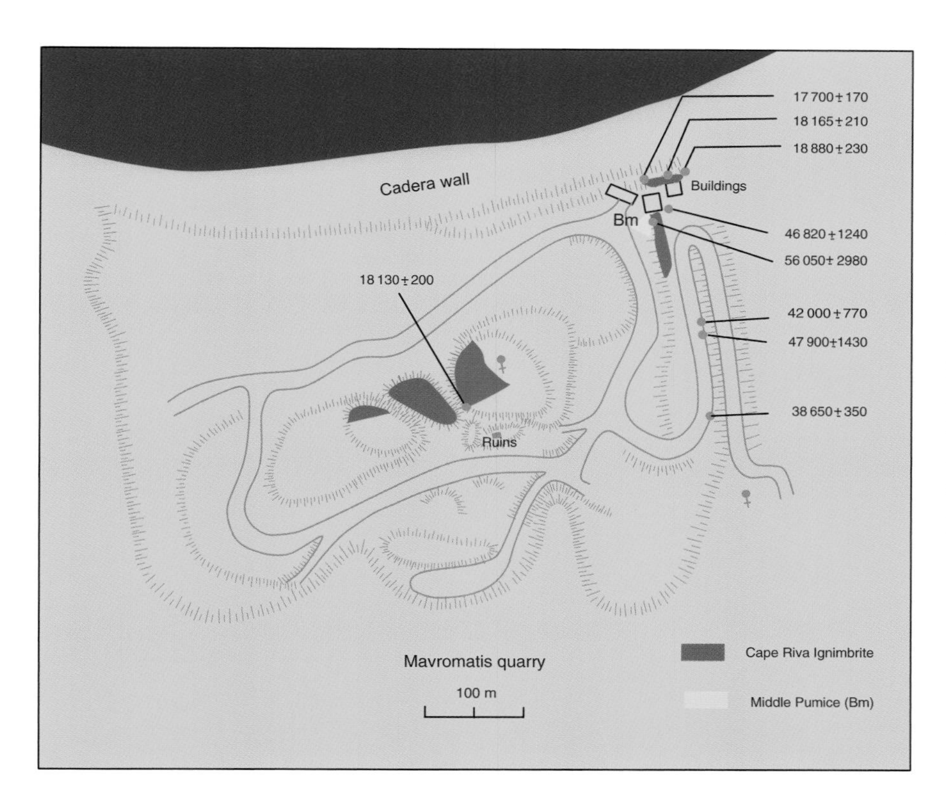

Figure 4.4 The Mavromatis quarry near Akrotiri contains vestiges of former vegetation such as tree molds and charcoal. The latter makes it possible to use the radiocarbon method to date the layers. Thus, the Middle Pumice layer (Bm) was dated to 56039 ± 2980 BP and charcoal in erosional channels gave ages of about 41 000 years BP. The Cape Riva ignimbrite contains about 50 tree molds with charcoal with ages of about 18 000 radiocarbon years or, by calibrating, 21 000 calendar years ago (see Chapter 7). After Knudsen (1998) and Lund (1998).

preserved only when thin-fluid lavas or fine-grained ash covered them. Even though the organic material or impressions of former plants can not be identified in many cases, one can usually date the charcoal that is occasionally left behind by the radiocarbon method.

On Santorini this method has been used to advantage for the Minoan eruption. Plant remains found in storage jars at Akrotiri were especially useful for dating. A detailed description of this method and its results are given in Chapter 7.

Tephrochronology

One of the most useful stratigraphic tools is tephrochronology, a method used in Denmark as early as the beginning of the twentieth century. The internationally renowned work of Thorarinsson (1944) on Iceland showed its importance. When an explosive eruption occurs, a synchronous ash fall (a tephra layer) is laid down over a wide region. Such an ash fall is a valuable time horizon that represents a very short span of geologic time. It can usually be found in all deposits, on land as well as under the sea. Thus, if one can determine the age of the ash in one place, the entire layer is then dated. It then becomes a red strand running through the story, tying together the stratigraphy of various regions.

Five tephra layers from explosive eruptions on Santorini can be used to establish time horizons in the Eastern Mediterranean: the two layers of the Lower Pumice, the Middle Pumice, the Cape Riva ignimbrite, and the Minoan ash fall.

The two layers of Lower Pumice (Bu) have been correlated with the V-1 ash of deep-sea sediments (McCoy, 1980, Keller, 1981; Vinci, 1985).

The Middle Pumice (Bm) is exposed in the caldera walls on Thera, where it can easily be followed for more than a kilometer. It has been correlated with the W-2 tephra in deep-sea sediments throughout much of the Mediterranean (Keller, 1981; Vinci, 1985). Friedrich and Velitzelos (1986) assigned it an age of about 60 000 years, but new radiocarbon data give an

Figure 4.5 The deposits of the third phase of the Minoan eruption are commonly eroded to bizarre forms, as shown by this example on northern Therasia. They contain numerous dark foreign rocks (xenoliths) derived from older lavas. In the background is the island of Thera with the town of Oia and the Megalo–Vouno complex on the right.

age of 56 050 ± 2980 years (Figs. 4.3 and 4.4), (Knudsen, 1998).

The Cape Riva ignimbrite forms a marker horizon about seven meters thick throughout much of Santorini. This layer was correlated with the U-2 tephra in deep-sea cores (Keller, 1981). Several ages have been determined by the radiocarbon method using fragments of charcoal (Friedrich & Pichler, 1976; Friedrich *et al.*, 1977; Eriksen *et al.*, 1990). Most measurements gave an age of 18 000 years BC which, when calibrated according to the calibration curve of Bard *et al.* (1990), gives an age of 21 000 calendar

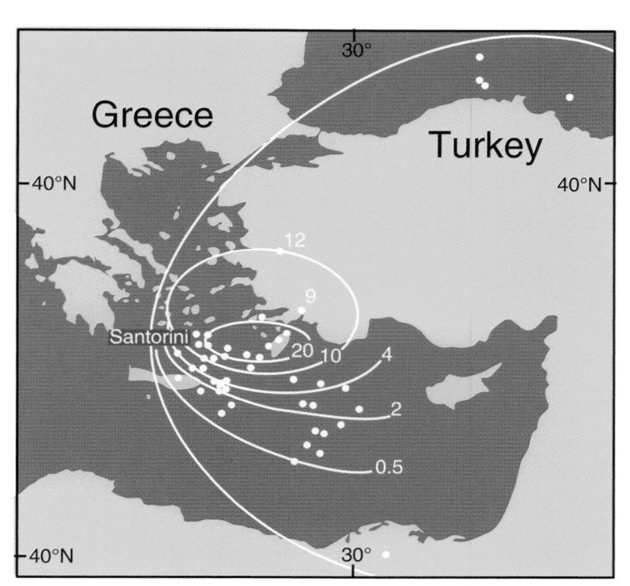

Figure 4.6 The ash-fan of the Minoan eruption is one of the most important marker horizons in the eastern Mediterranean, Anatolia and the Black Sea. The layer thins toward the east. The isopachs are in centimeters. (After McCoy and Heiken, 2000.)

years. This age can be considered reliable, since it was based on charcoal derived from many small trees that were covered by the ignimbrite. The extent of this ignimbrite has been described by Pichler and Kussmaul (1980) as well as by Druitt (1985).

The Upper Pumice (Bo) buried Bronze Age buildings on Santorini and the eruption spread an extensive ash fall over the southern Aegean Sea. This eruption is therefore an important time horizon for archaeologists as well as geologists (Figs. 4.5 and 4.6). While archaeologists estimated an age for the eruption of about 1500 BC on the basis of ceramic artifacts (Doumas, 1983), the radiocarbon age calibrated to the decay curve indicates a date of 1640 ± 7 years BC (see Chapter 7).

5

Plant remains and geological time

Fossil plants give us a glimpse into the vegetation of the past. They also reveal to us the climate that prevailed on Santorini and throughout the Aegean region. The climatic variations of the last ice age are also recorded in marine sediments around Santorini; they clearly show that there was never an ice sheet covering this region. During glacial periods there were only minor plants and grass, while during warm intervals trees grew on the islands. The climate of the Bronze Age resembled that of today.

Preservation of plant fossils in volcanic regions

About 18 000 years ago, when the last ice age reached its maximum, the climate of Santorini was cooler, just as it was in northwestern Europe, but it is certain that no ice sheet ever reached the islands. How do we know this? The only indications one can call upon to reconstruct the climate are plant fossils that one finds in the volcanic deposits. One might wonder why we find no animals here. It is logical that land animals seldom leave fossil remains in volcanic regions, because they are better able to avoid the threat of natural catastrophes. Plants, however, do not have this ability. Even when they are overwhelmed and buried by ash or lava, they are not totally destroyed; some trace is almost always left behind. They may leave cavities, impressions, or fragments of wood and charcoal (Fig. 5.1). A careful search often reveals the traces of former vegetation on the underside of a lava flow or beneath layers of ash (Figs. 5.2 and 5.3).

On Santorini, as in other volcanic regions, as loose slag and ash accumulate, plants grow preferentially in the erosion channels that develop on volcanic slopes. In these streams the plants are sheltered from wind and can find water. But when an eruption occurs this is certainly a major disadvantage, because lava flows or glowing avalanches tend to follow the existing drainage lines. Thus, the vegetation concentrated in

Figure 5.0 In this view of Millo Bay on Therasia, the red ignimbrite can be seen in the middle part of the photograph. It contains on its lower part charred plant remains that radiocarbon measurements have dated at 21 000 years BC.

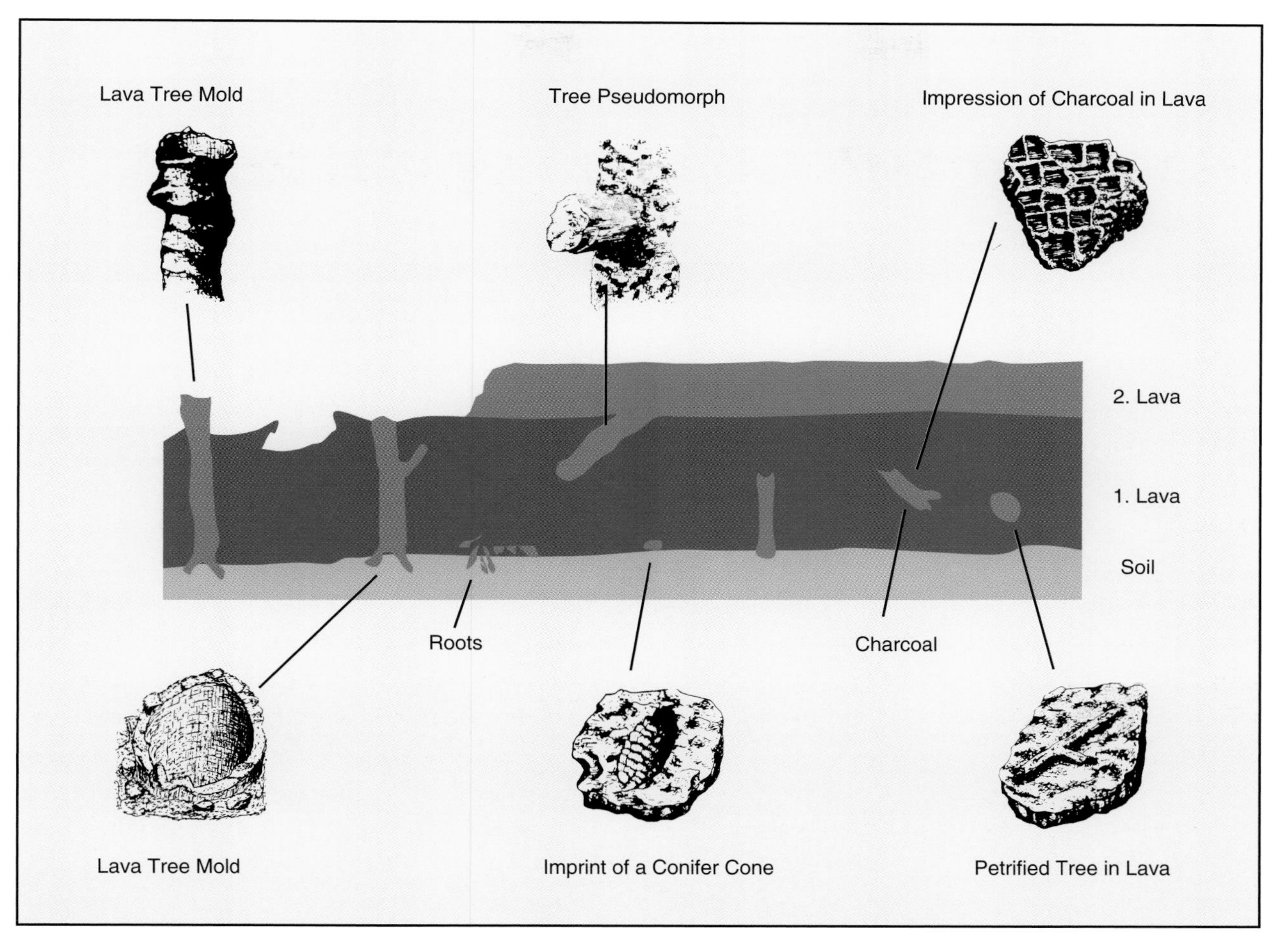

Figure 5.1 Schematic diagram showing how plant remains are preserved when covered by fluid lava. The Leidenfrost phenomenon plays an important role under these conditions. Just as a drop of water falling on a hot stove will not vaporize immediately but dances on the surface because an insulating layer of steam surrounds it, the mantle of steam surrounding wet vegetation forms a shield between it and the glowing lava. Thus the lava will cool around a damp tree and form a cast. As is commonly seen on the slopes of volcanoes, a freestanding lava mold is left when the lava drains away. These tree molds may contain charcoal or show casts of the bark on their inner sides. They may later be filled with new lava or deposits of secondary minerals making a cast of the original tree.

these channels is covered and incorporated into the debris. The degree to which traces of vegetation are preserved depends on several factors. The temperature of the overlying volcanic material is especially important as is the moisture content of the vegetation and the presence or absence of oxygen. With elevated temperatures and free access to oxygen, for example, the vegetation is burned and only in exceptionally favorable conditions can impressions and cavities be found in the host material (Fig. 5.4). If there is little oxygen and temperatures are below about 360 degrees Celsius, even organic material may be preserved.

The plant fossils of Santorini

The plant fossils of Santorini were first described by the French geologist Alfred Lacroix (1896) who, together with his colleague (and father-in-law) Ferdinand Fouqué, collected specimens from a quarry south of the town of Fira. It seems quite likely that fossils of this kind had already been recognized on the island, since both these geologists were told about

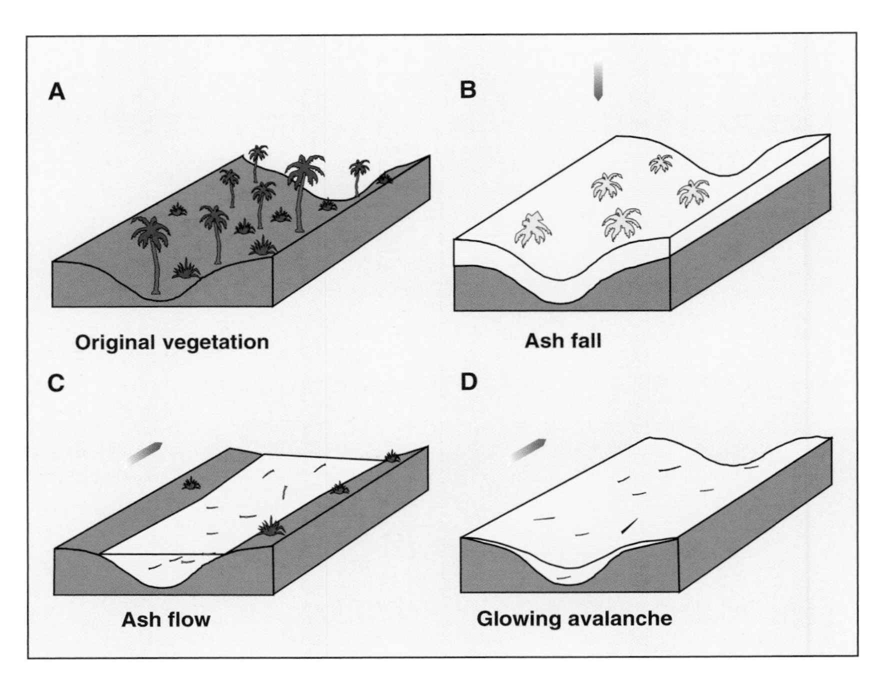

Figure 5.2 Preservation of plants in loose volcanic material depends on the temperature and direction of motion (arrows) of the imbedding material. When a valley (A) with vegetation is covered by a thin layer of ash (B) the vegetation remains in place and may even survive a light ash fall. In the case of ash flows, however, most of the vegetation is torn from its original site and oriented parallel to the flow (C). Depending on the moisture of the imbedding material, the vegetation may be partly or totally destroyed. Vegetation can also be moved from its original site and strongly abraded by a glowing avalanche (D).

Figure 5.3 In a deeply weathered tuff, calcified roots look like bleached bones. Because they are harder than the tuff, wind erosion makes them stand out on the surface. Such plant remains indicate a long interval of quiet in the sequence of eruptions during which vegetation could be re-established and the ash could be strongly weathered.

Figure 5.4 Remains of charcoal are very important for dating layers, but only when their age lies within the useful range of the radiocarbon method. It is sometimes even possible to identify in a thin section of coal the kind of plant one is dealing with. Finally, the presence of charcoal indicates that the temperature of the imbedding medium must have been greater than 360 degrees Celsius. The charcoal shown here is from an erosional channel that descends toward the present caldera. It is exposed in the Mavromatis quarry near Akrotiri (see Fig. 4.4). A radiocarbon measurement giving a date earlier than 42 000 years BC means that this minimum age can also be used to date the southern part of the caldera.

them at the Lazardist monastery. The fossils no doubt became known when pumice was being used to construct buildings in the vicinity of the town. According to Lacroix's account, the locality where he collected the plant fossils was 15 meters below the lava on which the town is built. This is more-or-less the same place where Julius Schuster (1936) obtained his collection.

The plants described in the following section were also collected at this same locality. Lacroix's fossil material is now stored in the Museum of Natural History in Paris, while those of Schuster can be found in the Natural History Museum of Berlin. There are other specimens in the Senckenberg Natural History Museum in Frankfurt am Main and in the collection of the Catholic church on Thera (Megaron Ghysi). In addition, a plate with impressions of olive leaves is stored in the Archaeological Museum on Fira.

Plant remains of the Fira Beds

The tamarisk tree (*Tamarix* sp.)

Until now only a few small remnants of tamarisk trees have been found on Santorini. They belong to the family of Tamaricacean trees commonly found today in the Mediterranean region. They are highly valued as shade trees, especially in coastal areas, and are still cultivated on the shores of Santorini for this purpose. At some time in the past tamarisks disappeared from Santorini. As Heldreich has noted (in Hiller von Gaertringen, 1902), they were reintroduced from neighboring islands by the Santorini scholar De Cigalle.

The lentisc plant (*Pistacia lentiscus* Linné)

I have been able to describe a few clusters of leaves in the tuffs of Fira and, with the help of the electron microscope, examine their cuticle (Friedrich, 1980). In this way it became clear that it really was *Pistacia lentiscus* Linné, which is still widespread in the Mediterranean region today (Fig. 5.5). On Santorini it can be found today on Palea Kameni (Hansen, 1971; Schmalfuss, 1991), and I have also found it on Thera in the area around Archangelos. *Pistacia* bushes, which are in the cashew family (Anacardiaceae), are also characteristic elements of the modern Mediterranean flora.

Figure 5.5 The upper photograph shows a pinnate leaf of *Pistacia lentiscus* Linné. A single leaf (below) that was part of such a compound group has a length of 18 millimeters.

The therebinth (*Pistacia therebintus* Linné)

Only a few fossil leaves of *Pistacia therebintus* have been found on the island of Thera. They are distinguished from the lentisc bush described above by their odd-pinnate, feathery leaves. This species also occurs in the modern Mediterranean region, where it is often encountered in association with *Pistacia lentiscus.*

The olive tree (*Olea europaea* Linné)

More than a hundred separate olive leaves have been found in the quarry south of Fira. Judging from the shape, venation and other anatomical details of the leaves there can be little doubt that the fossil trees were *Olea europaea* Linné. Under the microscope, one can see fine protective hairs on the underside of the leaves – identical to those of modern trees. They form a white, felt-like layer that serves to reflect sunlight and reduce evaporation from stomata in the underside of the leaf. The ancestral olive leaves of Santorini are the oldest known in the Mediterranean region. They are the best signs that this plant has been characteristic of the region for more than 50 000 years (Fig. 5.6) and was

Figure 5.6 Compressed leaves of *Olea europaea* with an age of about 50 000 years. They are the oldest known remains of olive trees in the Mediterranean. The leaves have a length of about 5 centimeters.

Figure 5.7 Fossil leaves of the olive tree (*Olea europaea* Linné) are often found in erosional channels of the Thera volcano. In this electron micrograph (top) one can still see remains of cocoons of the white olive fly (*Aleurolobus* (*Aleurodes*) *olivinius* Sylvestri). This parasite is also found on the leaves of modern olive trees on Crete (bottom). Thus, the dated fossil findings from Santorini show that this parasite–host relationship has lasted at least 50 000 years. The parasite has a size of 400 microns.

not first introduced by humans in historic times, as was once believed.

With the aid of the electron microscope, one can see parasites (*Aleurolobus* (*Aleurodes*) *olivinius* Sylvestri) on some of the fossil olive leaves (Fig. 5.7). They grew on the white felt-like layer of the olive. This parasite is also found on olive leaves today.

Fossil leaves with parasites of this kind are a notable example of the parallel evolution of plants and their parasites. Moreover, they are interesting in terms of the history of the region's vegetation, because they demonstrate that the relation between the host plant and the parasite that depends on it can persist for millions of years, as shown from an example from Iceland (Friedrich and Simonarsson, 1981).

The dwarf palm (*Chamaerops humilis* Linné)

Fragments of the dwarf palm are quite common in the quarries south of Fira and at Athinios (Karageorgis quarry). Their large palm-fronds with long, V-shaped spikes are found mostly in the characteristic 'double band', where an erosion channel was cut into the ash layer on the flanks of the volcano. *Chamaerops humilis* is quite typical of the Mediterranean region (Fig. 5.8). It was originally native to this region, but in historic times was wiped out in the western region by humans. In Greece, it is a popular ornamental plant. On Naxos the dwarf palm is still found growing wild (Friedrich and Velitzelos, 1986).

The date palm (*Phoenix theophrasti* Greuter)

Until now, fossil date palm has been identified on Santorini from only a few fragmentary impressions of fruit. Today, the date palm is found on Crete in the district of Vai and at Preveli, where it occurs with the endemic species *Phoenix theophrasti* Greuter, which is unique to that region, and date palms grow throughout most of Santorini (Fig. 5.9).

Figure 5.8 Imprints of leaves of the dwarf palm (*Chamaerops humilis* Linné) from the quarry south of Fira. The palm was common in the late Quaternary flora of Santorini, but today it is nearly extinct in the eastern Mediterranean. It is found in only a few places on the island of Naxos.

Figure 5.9 Date palms grow today on several places on Santorini where they find water and are protected from the winds, as they are in this part of the inner side of the caldera at Plaka. (Watercolor drawing by Barbara Gentikow, August 1991.)

In some cases it has been possible to establish the age of fossiliferous beds by means of radiocarbon dating of the plant material (Fig. 5.10). The age of one of the paleosols of Fira has been determined by radiocarbon dating of charred wood, which I found in the quarry at Fira. This measurement gave 54 250 ± 700 years BP (Friedrich and Velitzelos, 1986). A paleosol somewhat higher stratigraphically in the section at Fira has been given an age of about 50 000 years.

The way plant fossils usually occur in volcanic regions is clearly illustrated by an example from Santorini. In the quarry of Mavromatis near Akrotiri an ignimbrite fills an erosion channel where about 50 small, carbonized branches, about as thick as an arm, have been found at the base of the ignimbrite (Fig. 5.11).

A similar occurrence can be seen in the cliff above Millo Bay on Therasia, where the lower part of ignimbrites of the same age contains fragments of

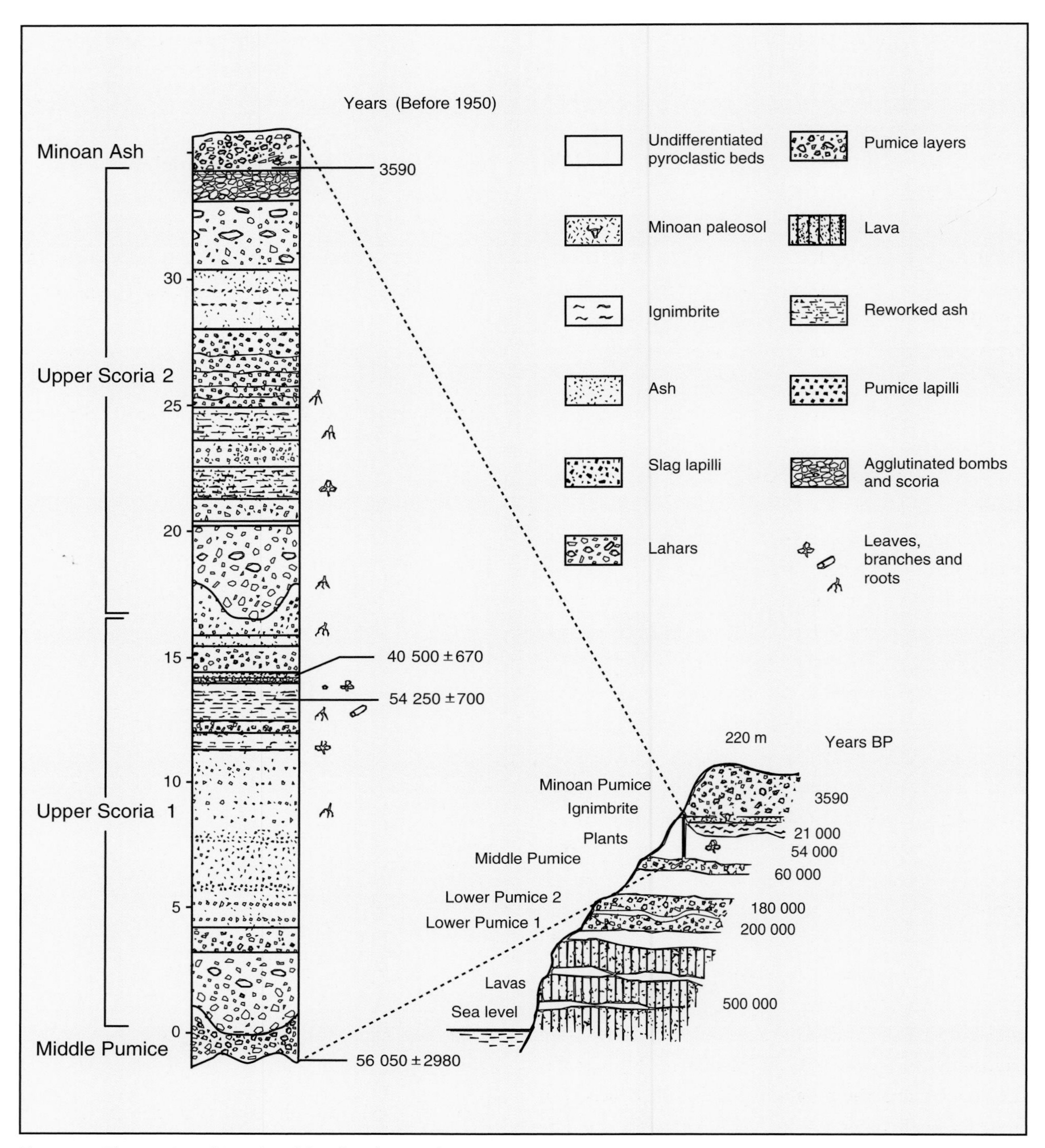

Figure 5.10 The stratigraphic order of the plant-bearing layers in the quarry south of Fira. Modified after Friedrich *et al.*, (1977).

stems about as thick as a finger. In both cases, the plant remains have been altered to charcoal and consist of only black powder in the cavities. For this reason, the carbonized plants are no longer identifiable but they can still be dated, for they have the same ages as the overlying Cape Riva ignimbrite. Radiocarbon dating of this material by different laboratories provides calibrated ages of about 21 000 years BP (Fig. 5.0).

Plants as indicators of climate

If fossil plants can be identified, they may provide direct evidence of the climate of the past. In the case of the fossils in the quarry south of Fira, this presents no problem. The plants that grew there about 50 000 years ago, and even the parasites that lived on them, still thrive in the Mediterranean region today. At Preveli on the southern coast of Crete there are places where the pistacia, tamarisk, and date palm, described above, are found in natural association (Friedrich, 1980). One can conclude from this that about 50 000 years ago the climate on Santorini resembled the modern one at Preveli on Crete.

This interpretation of past climatic conditions has been verified through comparisons with another region of Greece. Pollen-bearing horizons found in a peat bog at Tenagi–Phillippi in Macedonia (Wijmstra, 1969) could have the same age as some of those on Santorini. This correlation enables us to interpolate the climatic regime of the islands. During cool periods, an *Artemisia* steppe grew in Macedonia; in warmer times there were conifers and oaks. Further evidence of a warmer climate earlier than about 50 000 years ago comes from the evidence that trees grew in Macedonia, as well as on the islands of Santorini (Friedrich *et al.*, 1977).

Fossils are known to occur in the 21 000-year-old Riva ignimbrite where it is exposed in the Mavromatis quarry and at Cape Riva on Therasia. The plant fossils, which up to now have been found only in erosion channels filled with pyroclastic debris, support the impression that the vegetation at that time was confined to a few stream beds, and that in the late Quaternary the islands were almost barren of vegetation. Even if these plant fossils provided no indication of climate, it would still be possible to use indirect evidence. One can use radiocarbon techniques to determine the age of the ignimbrites and compare them with the fossils of similar deposits in neighboring regions, and thereby draw conclusions regarding the former climate on Santorini. Moreover, at Tenagi–Phillippi there are trees that were growing at the time the ignimbrite was erupted.

For Santorini one must not fail to take into account the special soil conditions on the slopes of a volcano where wind and rain played a decisive role for the flora. Wind and rain can of course rework porous volcanic debris, so that the growth of plants is hindered if not prevented completely. In addition, water percolates through the porous tuff deposits, sometimes descending to such depths that the roots of the plants are unable to reach it. It can create a desert in this way

Figure 5.11 The volcanologist Maurice Kraft placing his arm into one of the holes in the ignimbrite in Mavromatis quarry near Akrotiri. The molds were formed from small trees and shrubs growing in an erosional channel about 23 000 years ago. They were covered by a glowing avalanche and burned, and are now found as molds on the lower part of the ignimbrite. Some holes contained charcoal that could be dated by the radiocarbon method.

even though water is abundant deeper in the ground. For example, underground water of this kind is known to be present beneath the desert of central Iceland (Schwarzbach, 1963).

The flora of Santorini today

The geological conditions of the Tertiary and Quaternary had a decisive influence on the flora and fauna of the Cyclades region. In particular, the break up of the islands led to an interruption of migration routes and the development of endemic forms. It reduced the diversity of plants relative to that of the neighboring mainland of Peloponese and Anatolia. Greuter (1967) divided the flora of the Aegean into two geographic regions: the southern Aegean, and the 'cardaegaeis', or heart of the Aegean, to which belong the other Cycladean islands in the same group as Santorini. The southern Aegean region, including Crete, Carpathos, and Rhodes, formed a single floral zone that served as the connecting link between east and west and as a pathway between Europe and Asia. The central Aegean islands are different. Greuter writes that they 'correspond to isolated insular conditions in which a relatively large part of the old Tertiary flora survived the Pleistocene climatic changes, because they were protected from the competition of the cold-resistant species that spread on the mainland.'

Box 5.1 The history of research on the origin of the present-day flora of Santorini

Records show that the flora list on Santorini has grown from the first collections in the last century to more than 550 species. Interpretations of the flora, which at the beginning of this century were based on a species-poor assemblage, have changed since then, as more species have been collected and more has been learned about them.

In the years 1822 to 1881 Heldreich, when he was the first director of the botanical gardens at Athens, catalogued a plant assemblage for his work 'The Flora of the Island Thera', which was published in chapter 4 in Hiller von Gaertingen's monograph on Thera (Heldreich, 1899). His list contains 240 naturally occurring species and three naturalized plants. An addition after his death added to Heldreich's list 59 new spontaneous and naturalized species. The plants were collected in 1900 and 1901 by P. Wilski and local collaborators. This increased the known naturally occurring and naturalized species of Santorini to 300 (Heldreich 1902).

Vierhapper (1914 and 1919) examined the specimens collected by a Vienna University trip to Santorini and as a result he was able to increase the number of known species by 74. In addition, by 1971 the botanist Hansen of Copenhagen was able to increase the flora list by 116 new finds to a total of nearly 500 species. Thus, it is not surprising that the geographic interpretation of the flora has changed in the course of time. Before the work of Heldreich the flora was considered 'very poor', but later, even when 500 species were known, Hansen spoke of a scarcity of species which he attributed to the volcanic soil and the nearly complete lack of ground water, streams, and wetlands. He also remarked that the present vegetation is relatively recent, owing to the total destruction by the Minoan eruption, and observed that the island group has today nearly no endemic species. For comparison, he pointed to the number of species on the island of Milos, which also belongs to the Cyclades, where according to Wiedenbein (1988) 422 species are found.

Raus (1991), however, viewed this situation quite differently. He argued that 'one cannot consider the flora of Santorini as species-poor due to the volcanic eruption 3600 years ago', and that 'the present number of species is comparable with those of islands of similar size that have not had a total destruction of their population (for example, Kasos or Kithira).' In his opinion, the process of immigration of plant species to Santorini is continuing and in the long term one can expect a further increase in the number of species, because the island group is still far from being 'saturated.'

A list of the flora of Santorini is given in Appendix 3.

Santorini's modern flora has been strongly influenced by humans. In the uncultivated areas one finds plants similar to those on the neighboring volcanic island of Milos (Wiedenbein, 1988). A scrubland vegetation, known throughout the Mediterranean region as *maquis*, is composed primarily of leathery, broad-leaved evergreen shrubs or small trees. A poorer version of this type of vegetation, the garrigue, is found at the foot of the Archangelos Massif, on Palea Kameni and in a few nearshore zones on Akrotiri Peninsula.

Part 2

The Minoan eruption and its effects

6

The mechanism of the Minoan eruption

The catastrophic Minoan eruption wiped out the Bronze Age settlement on Santorini and buried it under a thick layer of ash. For more than 3600 years it lay hidden and forgotten under a shroud of white pumice. Not only Santorini, but also the neighboring islands and Anatolia, were affected by the powerful eruption. Ash, poisonous gases, floating pumice, and floods produced by tsunamis inflicted damage on large parts of the Aegean region.

The thick ash deposits left by the Minoan eruption have led to much discussion and speculation. Even today, after thorough investigations of the up to 60-meter thick pumice layer, there is still controversy over the puzzling observations related to its history. Having no historic examples of comparable volcanic catastrophes to judge by, the nature of the event can only be deduced from indirect evidence. Eruptions of such magnitude do not happen every day.

Box 6.1 The Tambora eruption of 1815

The eruption of Tambora volcano on the island of Sumatra in Indonesia was one of the stongest known. A clear acid signal was left in the Greenland ice following a delay of one year (Fig. B.6.1). A total of 150 to 180 cubic kilometers of ash and pumice were ejected, and a tsunami was generated by the sudden entrance of ash flows into the sea. Because the crater of Tambora lies 15 to 20 kilometers away from the sea at a height of 2850 meters, the tsunami could not have been triggered by collapse of the caldera or by phreatomagmatic explosions. The tsunami reached a height of four meters on the coast of Sumbawa and 2 meters on the coast of Java (Walker, 1973). About 90 000 people lost their lives as a direct result of the tsunami or indirectly from the famine that followed. The eruption had a global effect on the climate. The following year was called the 'year without a summer' (Stommel and Stommel, 1985).

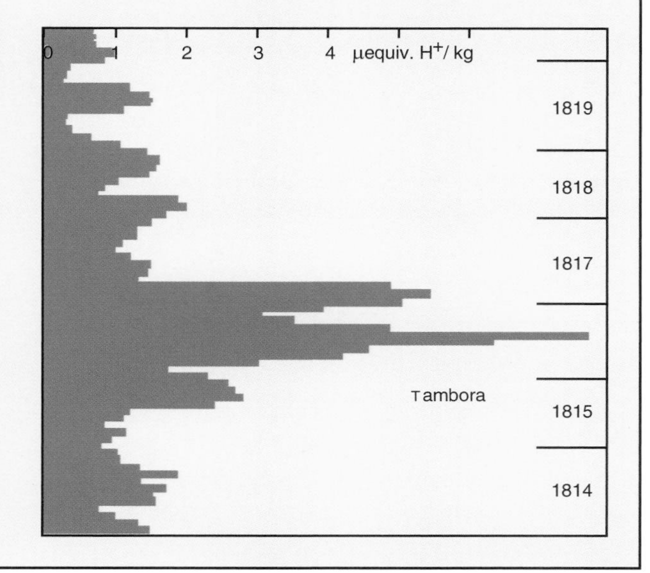

Figure B.6.1 It was very easy to recognize the acid signal from the Tambora eruption of 1815 in the bore-hole 'Dye 3' in the inland ice of Greenland. It was even possible to determine that the acid aerosols remained in the stratosphere for several months, as the acid signal did not appear in the ice until the snowfall of the year 1816. A similar acid signal was also left by the Minoan eruption. (Modified after Dansgaard and Hammer, 1981.) See also Fig. 7.4.

Figure 6.0 At Cape Alonaki south of Fira the three main units of Minoan pumice can be recognized from their distinct forms.

Since the Minoan eruption played such a definitive role, not only for the morphology of Santorini but also for the population living on the volcanic island, it is important that we try to understand the processes involved in this eruption. Only very recently has it been possible to construct a reliable picture of the full course of the eruption. In historic time, only the volcanoes Tambora (1815) (see Box 6.1) and Krakatau (1883) (see Box 6.2) had eruptions of similar magnitude. In the catalogue of known volcanic eruptions (Simkin *et al.*, 1981), the strengths of explosive eruptions have been classified according to a 'volcanic explosivity index' (VEI) (see Box 6.3) based on quantitative criteria. In this index, the Minoan eruption was a real 'heavy-weight', for its value of 6 is just below that of the Indonesian volcano Tambora (VEI = 7) and is

Box 6.2 The eruption of Krakatau of 1883

Situated in the Sunda strait between Java and Sumatra, Krakatau is part of the Indonesian volcanic arc, which is one of the most active in the world. When it erupted in 1883, its impact was worldwide (Fig. B.6.2). The atmospheric shock wave was registered in Potsdam, Germany, and the sound was heard on Madagascar, Sri Lanka, and Australia, at distances of up to 5000 kilometers away. The tsunamis it produced killed 36 000 people on the shores of Java and Sumatra and carried ships several kilometers onto the land. These same tsunamis circled the earth twice. Fine particles of volcanic glass and lithic debris were projected into the atmosphere where they remained for several months causing brilliant sunsets and unusual climatic effects. In places, the darkened sky was so red that the people believed a neighboring town was in flames. As in the case of Santorini, the amount of erupted material was enormous: Self and Rampino (1981) estimated it to have been 18 to 21 cubic kilometers or the equivalent of 9 to 10 cubic kilometers of dense magma.

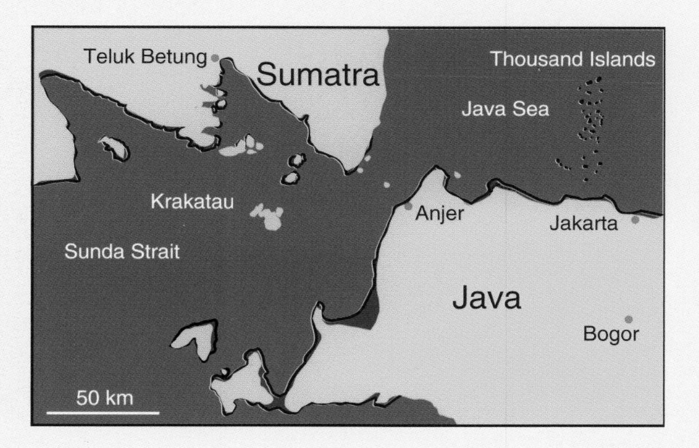

Our knowledge of the eruption is based mainly on the observations of the Dutch engineer, Verbeek (1886), who carried out a thorough investigation for the Netherlands government. He tried to explain the immense explosions and tsunamis as the result of interaction of the magma with water. This may have been a contributing factor, but certainly not the only one. Several alternative explanations have been proposed. For many years volcanologists accepted the idea that the collapse of the caldera triggered the tsunamis, but more recent work, such as that of Self and Rampino (1981), indicates that when the eruption column collapsed it produced dense pyroclastic flows that entered the water and generated the huge waves. This phenomenon is thought to have accompanied each of the four major explosions that occurred on the 27th of August, 1883. The timing of events suggests that the collapse of the caldera did not come until after the pyroclastic flows and tsunamis and could not, therefore, have been their main cause. The entry of sea water into the eruption crater was less important during the main phase of the eruption. It produced only minor phreatomagmatic explosions.

Figure B.6.2 In 1883 the volcano Krakatau, situated in the Sunda strait between Sumatra and Java, had a series of catastrophic eruptions. They triggered tsunamis (dark blue in the drawing) that penetrated deep into the coastal regions where they caused great damage. In all, 36 000 persons were killed by this catastrophe. Modified after Symons (1888).

equivalent to the eruption of Krakatau in 1883. Useful information can, of course, be obtained from accounts of weaker eruptions as well. In the 1965 eruption of Taal in the Philippines, for example, a phreatomagmatic eruption (see Box 6.4) was well documented for the first time (Moore, 1966). This new knowledge was subsequently used in the interpretation of several other eruptions, such as that of Laacher See in the Eifel (Germany). It also enabled Pichler (1973) to explain the puzzling undulating layers in the pumice on Santorini as base surge deposits. During the atomic bomb explosions at Bikini Atoll, similar ground waves were observed to emerge from the center of the explosion much as they probably did at Santorini.

Equally remarkable are the products of the eruption, especially the pumice. Because it is formed of a light, porous volcanic glass, it floated for months on the water and was easily carried for great distances by wind and waves. Its distribution has played an impor-

Box 6.3 Volcanic explosivity index

The VEI (volcanic explosivity index) uses a scale with eight steps as a measure of the strength of explosive volcanic eruptions. Among the parameters used to estimate the magnitude of an eruption are the height of the eruption column and the volume of ejecta. The catalog *Volcanoes of the World*, published in 1981 by the Smithsonian Institution of Washington, DC, includes all known historic and pre-historic eruptions on earth.

Box 6.4

Phreatomagmatic explosions

The interaction of molten magma with water creates a violent eruption due to the accelerated vaporization of water to steam. 'Base surges' are formed when the expanding water vapor moves outward from the vent at high velocities. In some instances, these surges form concentric rings.

Plinian eruptions

A plinian eruption sends a gas-rich column with huge amounts of loose volcanic ejecta to great heights above the volcano. The name comes from Pliny the Elder who lost his life in the 79 AD eruption of Vesuvius that buried the towns of Pompeii and Herculaneum.

Volcanic bombs

Large pieces of ejecta thrown out of a volcano are referred to as bombs if they were molten and became rounded in flight. Solid fragments of the same size but with angular shapes are called blocks. An indentation where a bomb has hit the ground is a 'bomb sag'.

tant role in reconstructing the ash-fan of the Minoan eruption. While the windborne part of the pumice was carried to the east, the currents of the sea carried another part in a southeasterly direction. Eventually, the part of the eruptive material that reached the stratosphere was carried in yet another direction, as will be discussed more fully later.

Was there warning of the eruption?

Clear indications of premonitory activity have been found in the excavations at Akrotiri. There is good evidence here of earthquakes which forced the inhabitants of the settlement to leave their homes: broken stair steps, collapsed walls, houses reduced to ruins, and heaps of debris gathered by the inhabitants have been found beneath the first deposits of the Minoan eruption (Fig. 6.1). The inhabitants were certainly given warning of an impending eruption, for they had time to remove food and valuables from the ruins. In one place, for example, they removed a set of three beds from the ruins and had them placed one atop the other with their posts upright (Fig. 6.2). Few valuables have been found in the excavations, and until today no human skeletons have been uncovered at

Figure 6.1 A broken staircase, ruined houses, and heaps of debris in the excavation of Akrotiri show that earthquakes struck the settlement before it was covered by the masses of ash from the Minoan eruption.

Figure 6.2 The inhabitants tried to remove some valuables from their destroyed houses, as in this set of three beds which they had placed one atop the other with their posts upright. However, earthquakes or other warnings of the impending eruption forced them to leave them behind. The original wood of the beds is long since gone but it left cavities in the pumice that one can fill with plaster and make castings of their form (See also Fig 9.1). Drawn after a photo by Sigrid Hedegaard, 1997.

Akrotiri. Nevertheless, as the uncovered frescoes clearly show, the inhabitants possessed weapons, such as spears and daggers, as well as gold ornaments. So it is probable that they were able to take away most of their valuables and foodstuffs.

A layer of fine ash that Doumas (1974) described from the excavations of the basal part of the deposits at Akrotiri can also be interpreted as an additional warning (Fig. 6.3). The ash, which has a thickness of three centimeters at Akrotiri, must have been laid down during the first stage of the eruption. Later studies by the two American volcanologists Heiken and McCoy (1984) showed that, in the Akrotiri region, this ash layer consists of up to 8 centimeters of the finest pumice particles. It is interpreted as the result of phreatomagmatic explosions during the opening phase of the Minoan eruption and was laid down in the form of fine water-quenched particles in the lowermost levels of the Minoan pumice on Akrotiri peninsula. These explosions were probably strong enough to be an effective warning to the inhabitants, assuming they did not come too late for the population to flee.

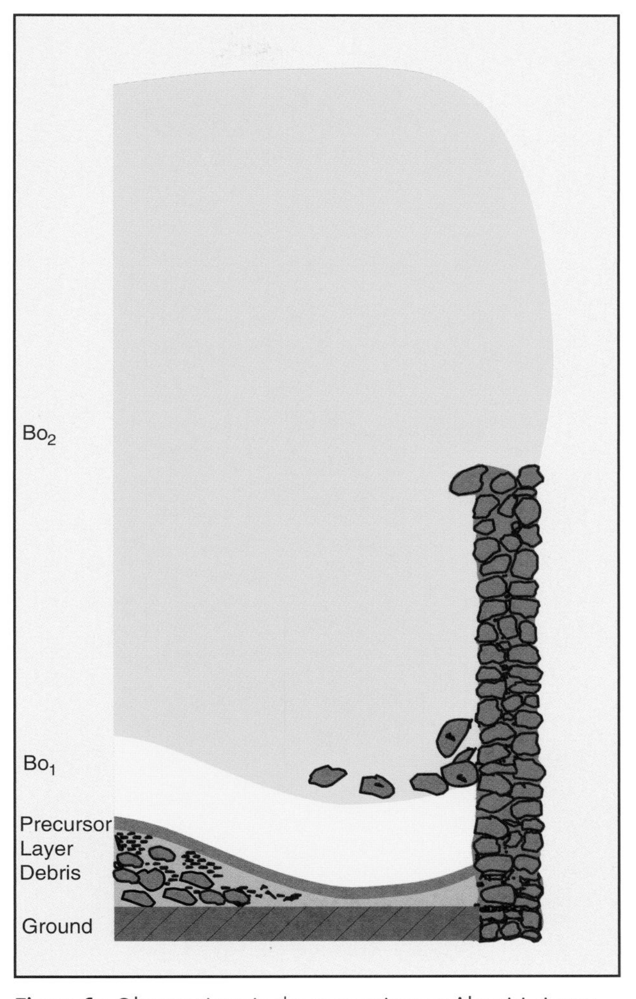

Figure 6.3 Observations in the excavations at Akrotiri give us important information about the mechanism of the Minoan eruption. At the bottom of the profile one sees pieces of debris that were covered by about 3 cm of pumice dust deposited before the first major eruptive phase. It is overlain by the units Bo_1, and Bo_2, which caused different types of damage to the walls of the houses. Schematic sketch partly after Doumas (1974, fig. 1).

Analyses of the descriptions of witnesses to similar historic events have shown that such warnings of eruptions are rarely recorded in folklore. It can be taken as certain that there were premonitory events prior to the Minoan eruption similar to those that formed the Kameni Islands in historic times (see Chapter 12). We know, for example, that a new island had appeared in the water-filled caldera as it did before the eruption of Nea Kameni in the year 1707. A brief but conspicuous sign had been given: blocks of stone were ejected with deafening explosions that echoed off the caldera walls. It is also possible that fissures were formed in the ground, as they have been seen to do in later eruptions and earthquakes at Santorini.

The first (plinian) phase

The first layer to be deposited was the 'Rose Pumice' (Neumann van Padang, 1936). It has a maximum thickness of seven meters in the region of Fira and thins rapidly toward the north, south, and west. At Cape Akrotiri, it is only 20 cm thick (Fig. 6.4); on Therasia to the west, it is 30 cm thick, and at Oia in the north only 50 cm. From these thicknesses it has been determined that the main wind direction during this phase of eruption was toward the east. Pichler and Friedrich (1980) have estimated the volume of this pumiceous material to have been 1.4 cubic kilometers, and calculations by Pyle (1990) show that the eruption column from which it fell reached a height of 36 to 38 kilometers (Fig 6.5). This is in agreement with earlier calculations of Wilson (1980) to the effect that the eruption column of the plinian phase reached the stratosphere and dust and gases as well as aerosols were spread through the atmosphere of the entire Northern Hemisphere (Fig. 6.6). This material was carried at least as far as Greenland (Chapter 7). According to the calculations of Sigurdsson *et al.* (1990), the amount of sulfuric acid from the eruption was greater than that of the eruption of Krakatau in 1883, which means that its traces can easily be detected in the Greenland ice cap.

The first explosive phase was very strong. As the reconstruction from the deposits shows, it began in a part of the ring-island where rising magma did not come into contact with sea water. The location of this eruptive vent is well established from various observations, such as the thickness and grain size of the eruptive products. It must have been in the region of the present Kameni Islands (Pichler and Kussmaul, 1972; Bond and Sparks, 1976). Our investigation (Friedrich *et al.*, 1988) showed that a large part of the caldera was at that time filled with sea water, but the products of the first phase show no signs of magma–water interaction, so the location of the caldera-forming eruption was most probably on an island. This hypothetical island has been called the Pre-Kameni island (Friedrich *et al.*, 1988; Eriksen *et al.*, 1990).

One can further conclude that the pumice-bearing first phase was ejected to a great height and that most of it fell back on the volcano and surrounding sea. Sea currents carried the floating pumice over a wide region, while fine ash was carried eastward by the wind as far as Anatolia, where it can be recognized today in

Figure 6.4 On the island of Thera the thickness of the Minoan ash diminishes rapidly to all sides except to the East. Near the lighthouse on Akrotiri peninsula the air-fall deposits measure only about 20 cm, as can be seen near the feet of Alexander McBirney, who is shown here as a scale.

Figure 6.5 The plinian phase of the eruption started within the caldera, where the Pre-Kameni Island was situated at that time. The eruption column in this phase is thought to have reached a height of about 36 to 38 kilometers. Drawing by Andreas Friedrich.

the sediments (McCoy and Heiken, in press). Pumice was washed up on the shorelines of the Mediterranean. Pichler and Friedrich (1980) speculate that this phase lasted only a few hours.

The second (base surge) phase

On Santorini the deposits of the second phase are easily distinguished from those of other phases due to their large undulating layers, some of which have wavelengths of up to ten meters. Owing to the fine grain size of the pumice, it can easily be recognized even from a great distance.

Volcanologists are now in agreement about the origin of this unit: in the second phase of the eruption, the eruptive mechanism changed completely. The

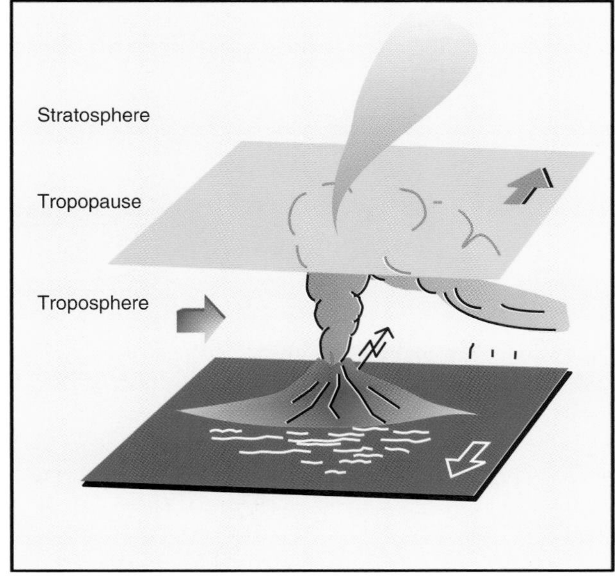

Figure 6.6 Schematic depiction of the different transport directions during the Minoan eruption. Sea currents carried the floating pumice to the south, while the wind carried another part to the southeast, and the jet stream in the stratosphere transported aerosols and very small particles to the north.

feeding vent of the Pre-Kameni Island had evidently widened, so that its surroundings broke down and cracks allowed sea water to enter. The coming together of sea water and fluid magma led to especially violent phreatomagmatic reactions. The magma was torn into small particles, which were surrounded by a thin layer of expanding steam. Clouds of ash suspended in steam spread outward from the eruption center in explosive rings and filled the entire caldera. They ascended the caldera walls, swept over the lower parts of the rim, and flowed down the outer slopes of the volcano. Observations at Taal Volcano in the Philippines showed that such flows of ash and steam can reach velocities of about 200 km per hour (Moore, 1966). Similar phenomena were observed at the atomic bomb tests at Bikini Atoll.

Figure 6.7 As seen in the light, fine-grained band in the upper part of the photograph the transition of the Bo_1 (ash-fall) to the Bo_2 (base surge) was not abrupt. The latter was formed when the magma came in contact with sea water. In this phase huge blocks of lava were thrown out (upper part of the picture). This profile is situated on the slope of Profitis Elias, southeast of the town of Pirgos about seven kilometers from the eruption site at a height of about 280 meters above sea level. At this elevation the thicknesses of the Bo_2 and Bo_3 layers are already strongly reduced, because they were formed by lateral flows that failed to reach the highest elevations. The person used for the scale is the Danish volcanologist Arne Noe-Nygaard. Photo by Erik Schou Jensen.

On Thera, the deposits of the second phase contain in their lower parts two plinian layers about 20 cm thick. They indicate that the passage from one eruptive phase to another went in pulses in which it was still possible for a phreatomagmatic phase to have two small eruption columns (Fig. 6.7). The magma–water contact then must have dropped to a greater depth below the surface of the sea and the overlying mass of water hindered formation of a tall eruption column.

According to calculations of Pichler and Friedrich (1980), about two cubic kilometers of pumice were deposited on Santorini in the second phase. These deposits have large wave-like structures that were a puzzle for geologists for a long time, until they were finally connected to the eruptive mechanism identified during the eruption of Taal Volcano in the Philippines in 1965.

On Thera, one sees that the base surge deposits have a very distinctive distribution. At Profitis Elias, for example, they are found only at elevations below about 350 meters. A similar situation is also seen on the flanks of Megalo Vouno Volcano. One can conclude, therefore, that during this phase the horizontally directed cloud flowed outward from an eruption center and over low parts of the caldera rim, sparing only the highest peaks. It is clear, therefore, that these are surge deposits and not the kind of ash layers resulting from the collapse of an eruption column from heights of up to 38 kilometers. If it had been the latter, the deposits should, of course, be found everywhere.

In both the first and second eruptive phases, huge pieces of lava were torn sporadically from the walls of the feeding vent and thrown out as blocks. They are clearly visible in the ash layers. On the caldera wall they can be seen forming 'bomb sags' in the Bo_1 layer

Figure 6.8 During the second phase of the eruption huge blocks were torn from the side wall of the vent and thrown out. Such a block is visible above the two people near the base of the cliff. If one studies the 'bomb sags' of these blocks one can often get information about the eruption point and the flight path of the block. Such measurements clearly indicate that the vent was situated in the present-day caldera. The diagram to the right shows the three phases of the Minoan eruption that produced the tuff units in the photo.

(Fig. 6.8). Direction indicators in these bomb sags in the caldera rim at Cape Athinios and Cape Alonaki show that the blocks were projected along ballistic trajectories from an eruption center within the caldera. One finds room-size blocks in exposures south of Fira that were certainly ejected during this eruptive phase. Blocks more than a meter in diameter also reached the Bronze Age dwellings at Akrotiri about ten kilometers from the vent. Even at this distance, they were still able to shatter stone walls (Fig. 6.9).

The third phase

The tephra of the third phase is easily recognizable in the caldera wall, even from great distances. It is distinguished from the other layers mainly by the large numbers of dark fragments it contains. The

Figure 6.9 During the second phase of the eruption huge blocks of lava landed in the settlement at Akrotiri and demolished the houses. One of these blocks is visible just behind the smashed wall (arrow).

fragments were rounded in the eruption column and mixed with the pumice. Most of these well-rounded, dark, glassy blocks are very similar to the lavas of Skaros Volcano. For the most part, these blocks come from lavas of the Pre-Kameni Island, which were thrown out, together with pumice and sea water, as the vent became wider during this phase. In this case too, the origin of the layer can be understood from its deposits. As the vent grew wider more of the crater wall gave way. It has also been deduced that in this phase the ash column did not reach the great heights attained in the first phase. Instead, turbulent clouds of ash and hot gas were directed laterally outward at low angles. These pyroclastic flows resembled a pot of overheated milk that boils over on the stove. As in the second phase, once they had passed over the caldera wall they flowed down the outer slopes of the volcano.

In the deposits of this phase, one finds a variety of types of xenoliths, including light-colored blocks of lime (Friedrich *et al.*, 1988; Eriksen *et al.*, 1990), the origin and occurrence of which are discussed more fully in Chapter 10. The large number of fragments of non-volcanic rock in the third phase of the eruption indicates that the rapid evacuation of the magma chamber led to a collapse of large parts of the low levels of the volcanic edifice. It produced the large northern basin of the present caldera and deepened the earlier-formed southern basins. The latter were partly refilled with pyroclastic ejecta during later eruptions.

Was there a fourth phase of the eruption?

The profiles in the quarries south of Fira and at Athinios and Akrotiri clearly reveal volcanic deposits that are especially rich in fragments of lava and overlie the pyroclastic flows of the third phase. Their dark color makes them easy to recognize. Experts disagree, however, as to whether they are primary products of an eruption or reworked material. While Pichler and Kussmaul (1980) mapped them as reworked deposits, Heiken and McCoy (1984) considered them lag deposits left by the eruption column in the closing phase of the eruption. They have the following notable features:

- The deposits consist of lithic fragments with little associated pumice.
- In places, the underlying layer (Bo_3) is clearly eroded.
- The deposits are found only in the lowest parts of the caldera rim as, for example, east of the village of Akrotiri and in the northeastern part of Thera.

The last of these observations argues against a primary deposit and in favor of reworking. One can imagine that collapse of the northern basin of the caldera caused a sudden inflow of the surrounding water which, on rebounding, went over the low parts of the caldera wall washing away some of the pumice laid down earlier. In this way, the light pumiceous material would have been removed, leaving behind the heavier lithic debris. We know from historical sources that in the year 1650 AD a tsunami hit the eastern coast of Thera in the area of Cape Kolumbo, Perissa, Kamari and Akrotiri. The effects on these regions are mentioned in the reports of eyewitnesses. As we shall see in Chapters 9 and 10, this tsunami could have contributed to the reworking of the Minoan deposits. Finally, wind and water erosion and human activity have strongly modified the uppermost part of these deposits by increasing the proportions of lithic components. The roots of grapevines penetrate the soil to depths of as much as eight meters, and when a vineyard has become too stony the farmers plow the field until they reach the soft pumice again. I am inclined, therefore, to favor the opinion that the 'fourth phase' is just a product of several of these secondary processes.

The effects of the Minoan eruption

Geological evidence on the remains of the ring-island

The Minoan ash layer, which was up to 60 meters thick, blanketed the former island ring. The evidence for the thickness of this layer in the literature is very uncertain. It depends on the assumption that the original 60-meter thickness is no longer preserved. Only a few of the caldera-wall sections still have their original thickness, because large amounts have been quarried and carried away. Fouqué (1879) gave for the region of Oia on Thera a total thickness of up to 60 meters, but by the 1970s (100 years later) the maximum thickness in a quarry in that region was only 35 to 40 meters. The pumice layer is not present everywhere on the islands; topography has had a great influence on its present distribution. Very little was deposited at the higher elevations, and where slopes were very steep it was removed by erosion. Tsunamis could have washed away some of the pumice at low elevations near the shores.

The scarcity of pumice at the highest elevations on Profitis Elias seems to indicate that such places remained unaffected by the eruption. Appearances are deceptive, however. At least the first phase of the eruption covered even the highest hills with a meter-thick layer of pumice which, because of the exposed, steep slopes, was quickly stripped by erosion. We find traces of ejecta from the first phase in crevasses on the high surfaces or as exotic blocks left on the surface when the lighter pumice with which it was mixed was eroded away.

The original form of the island was also altered by the eruption. It seems that much of what was lost from the central part of the ring-island was added to the outer margins. In particular, the eastern side of Thera was considerably widened by debris washed from the rim and deposited on the alluvial plain (Fig. 6.10). The former island of Monolithos was joined to Thera, whereas the Pre-Kameni Island inside the caldera disappeared completely. Its location is now the site of the northern basin, the bottom of which –

immediately after the eruption – had a depth of 500 meters below sea level. In addition, deep channels were cut into the original horseshoe-shaped island, leaving the three segments of Therasia, Thera, and Aspronisi.

Effects on flora and fauna

These eruptions must have had a profound effect on the vegetation of the older island. Minoan ash deposits destroyed almost everything. The only places where plants were able to survive were at high elevations,

Figure 6.10 The Minoan eruption added the broad, flat area in the right side of the photo on the eastern shore of the Bronze Age island. In the foreground the reddish limestone of Echendra is visible. Before the eruption, this area was accessible by boat.

such as Profitis Elias and the ridge of Platinamos, and on steep slopes in the wind shadow of the eruption. The botanist Raus (1991) considers it most likely that 'a few forms living in crevices, such as Chasmophytae or onion plants, were sheltered from the direct heat and mechanical effect of the eruption and could have successfully survived a temporary, limited cover with volcanic ash (e.g. Asparagus, Muscari or Scorzonera).'

The same was true for animal life, which had retreated to the upper elevations of the Elias massif. A few small creatures, such as snails, lizards, snakes and insects, survived the catastrophe in such places, while all other land animals were destroyed. Up to now, we have little evidence from the excavations of Akrotiri as to what the humans experienced, or that any perished there. It is more likely that the volcano provided a warning and they were able to leave the island in boats. But one cannot exclude the possibility that future excavations here will still turn up surprising discoveries. In Italy, at the large settlement of Herculaneum a pier has recently been found where the people crowded in a vain attempt to find shelter from the eruption of Vesuvius (79 AD) and were destroyed by the ash.

What do the neighboring islands show?

Pumice and ash must certainly have covered the nearby islands as well. The strong explosive eruption with its ash fall and noxious gases probably had a catastrophic effect on the entire surroundings of Santorini. The islands of Anaphe and Rhodes lying east of Santorini must have been subjected to a rain of ash, which was carried mainly in that direction. The Minoan ash layer can be recognized in many places on Rhodes (Keller, 1980; Doumas and Papazoglou, 1977).

Ultimately, it reaches as far as Anatolia. The huge mass of pumice undoubtedly covered the surface of the sea over a wide region and was washed up at higher levels on the shores by the tsunamis triggered by earthquakes. The causal relation of tsunamis was demonstrated by the earthquake of 9 July 1956, when the tides on the island of Ios reached a height of 25 meters (see Chapter 13). On most of the shores of the surrounding part of the Aegean, lumps of pumice have been found that clearly had drifted there on the surface of the water (Fig. 6.11). Pumice was also found on the northern coast of Crete and on the shores of Anaphe, Limnos, Paros, Samothrace, Cyprus, and even Israel (Francaviglia, 1990). It was also observed on the Nile delta (Stanley and Sheng, 1986).

During the transition from the first to the second phase, the eruption column that was directed to the east suddenly collapsed and generated strong tsunamis when the enormous mass of material suddenly entered the sea east of Santorini. Similar tsunamis were produced by the 1883 eruption of Krakatau. The deep basin in the northern part of the caldera indicates that tsunamis could also have been produced by collapse of the roof of the magma chamber during the third phase of the eruption. Floating pumice must certainly have hindered shipping and fishing for many months throughout much of the Aegean. Crops were probably destroyed in exposed coastal regions of Santorini and nearby islands. In addition, fine ash particles were carried into the stratosphere where they intercepted part of the sun's radiation and altered the climate on a global scale. This too would have contributed to widespread crop failures and famine.

Was Crete hit by the eruption?

For decades, this question has been the concern of scientists, especially in the light of the earlier discovery of the Minoan culture at Knossos on Crete by Sir Arthur Evans at the beginning of this century. Recognizing that a destruction of Knossos occurred at about the same time (Late Minoan IA), as was indicated by the style of ceramics found on Santorini in the 1860s, Evans began to speculate about a possible connection between the eruption of Santorini and the destruction of Knossos. Spyridon Marinatos (1939), who also worked on Crete, generated much

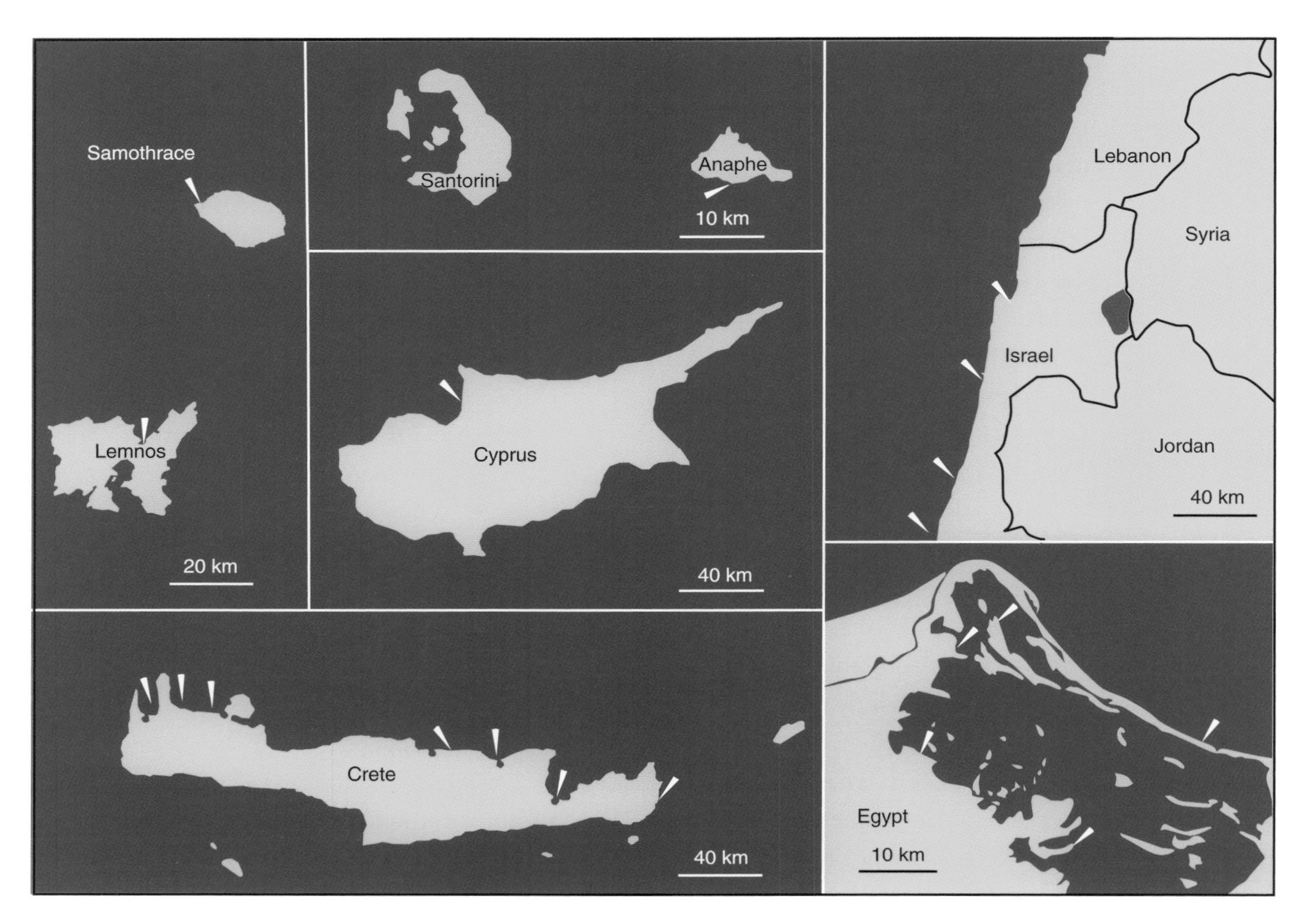

Figure 6.11 Pumice pebbles from the Minoan eruption (arrows) are found today on several beaches in the Mediterranean (Francaviglia 1990). They reached these distant shores by floating on the surface of the sea. It has also been possible to trace fragments of pumice from the eruption in the sediments of the Nile Delta (lower right) (Stanley 1986).

discussion by proposing the hypothesis that the demise of the Minoan culture on Crete was a consequence of the eruption of Santorini. According to his explanation, the eruption was associated with strong earthquakes, which could have caused great damage to the Minoan settlements on Crete. Moreover, the settlements on the northern coast of Crete would have been totally devastated by tsunamis triggered by the eruption. He cited as a documented analogy the eruption of Krakatau in 1883, which had many similarities to that of Santorini.

Today it is thought that Crete was spared the most severe effects of the widespread ash fall. Only the eastern tip of the island was covered with a few centimeters of ash. The possibility that Crete was hit by tsunamis triggered by the eruption of Santorini cannot, of course, be ruled out. High floods would have overwhelmed at least the north-coast harbor at Amnissos, one of the ports serving Knossos. It is unlikely, however, that the palace of Knossos was destroyed by tsunamis, because it has an elevation of about 60 meters above the sea. Moreover, the small nearby island of Dia would have intercepted any tsunamis that reached the north coast of Crete. Nevertheless, they could have damaged the coastal settlements and the ships anchored there would have been destroyed. For the destruction of towns on Crete, however, one must call upon a much more devastating event, such as powerful earthquakes and fires (Pichler and Schiering, 1977). Some scholars have concluded that the Mycenaeans of the mainland of Greece had conquered the weakened settlements, but internal political problems could also have brought on the collapse of the Minoan civilization.

Tsunamis also have unexpected secondary effects. As Wiedenbein (1991) showed, they can contribute to the dispersion of flora and fauna in a region like the Cyclades, where animals and plants washed into the sea were carried on floating pumice and by this means of dispersal migrated into new regions.

Climatic deterioration in China

According to Pang *et al.* (1989) the effects of the Minoan eruption were also felt in China. Ancient historical accounts mention severe deteriorations of climatic and hydrologic conditions in the seventeenth century BC. Several effects could have been caused by the eruption. Dry fogs and dim sun were recorded during the reign of King Chieh. The weather was colder and more irregular. Crop failures brought famine, and heavy rains caused flooding and destruction on towns. The floods were followed by seven years of severe drought. Pang and his colleagues dated these events using archaeologically verified pre-dynastic Shang and Western Zhou royal genealogies, which were calibrated by astronomical dates such as solar or lunar eclipses. According to their investigations, the eruption occurred 24 generations before 841 BC, which would place it during the reign of King Chieh in 1600 plus or minus 30 years BC.

Comparable eruptions

The catastrophic effects of the historic eruptions of Tambora (1815) (see Box 6.1) on the island of Sumatra and Krakatau (1883) (see Box 6.2) in the Sunda Strait have clearly shown how much damage must have been caused by the Minoan eruption. An event as great as the Minoan eruption must have left lasting marks in its nearby and distant surroundings. For example, ash falls and earthquakes – with or without tsunamis – are known to have been caused by caldera collapse or by the fallback of an eruption column.

7

When did the catastrophe occur?

The powerful Minoan eruption produced a synchronous ash layer over a great part of the Mediterranean region. Geologists and archaeologists are very interested in dating this unique marker horizon. If its age were known well enough, one could use it as a guide for the stratigraphy of the entire region.

One would think that, with all the methods of modern archaeology and natural science now at our disposal, it would be a simple matter to determine the timing of a natural catastrophe as great as the Minoan eruption. Unfortunately, that is not the case. In fact, we can determine the time of the eruption only within a range of a few decades, which is much less precise than is needed for an event of this importance. An adequate dating of the Minoan eruption would be a key to the chronology of the Bronze Age throughout much of the Mediterranean region. If the age of the pumice layer were known well enough, the age of all objects that it covers could also be adequately defined.

It is possible that in the near future a new technique for measuring ages will be developed. One might be able to use, for example, the tiny particles of volcanic glass that were carried by currents in the stratosphere to the inland ice sheet of Greenland where they are still detectable in deeply buried levels of the ice.

The methods by which one can determine the age of this eruption can by divided into two groups, one that gives relative ages and the other that gives absolute ages. The basic principles of both are employed in archaeology as well as in the earth sciences. Of course, the accuracy required by archaeologists is many times greater than that needed by geologists. Moreover, the periods of time that are of interest to archaeologists are relatively recent.

All possible methods have been used to date the volcanic catastrophe. Using relative methods one can compare an object with other objects, the age of which has already been determined, provided, of course, that one has adequate material of known age. In archaeology as well as in the earth sciences the ages of fossils are commonly determined by comparing them with index fossils, the age of which is already known. In making such comparisons there is always the danger of getting into a vicious circle, because, once an incorrect age is introduced into the literature, it is difficult to eliminate it later.

Archaeological estimates of age

Archaeological discoveries at Akrotiri enable one to compare the style of objects with those of articles found in other excavations in the Aegean region. In particular, Crete obviously had much relevance to Thera during the Bronze Age and has proved to be of great importance for temporal correlation and dating. It is possible to learn much about the commerce with other Minoan settlements from pictures, such as those of frescoes, or from decorated ceramics that provide important clues for the comparative dating. In addition, links to Egypt are indicated through discoveries such as the cartouche of King Khyan (of the Hyksos period) on an alabaster lid from Knossos on Crete (Fig. 7.1). Further reference points are provided by the depiction at Thebes (Egypt) of gift bearers from the Aegean region (about which more will be said in Chapter 11). Until only a few years ago, comparisons of this kind gave the archaeologist temporal reference points indicating that the Minoan eruption occurred between about 1500 and 1550 BC. In 1987, however, increasing numbers of radiocarbon age determinations indicated a greater age (1600 to 1650 BC) for the eruption and the archaeological chronologists began to adjust to this (Betancourt, 1987).

Meanwhile, new discoveries adding to the ambiguities of the dating appeared and extended new chrono-

Figure 7.0 The caldera wall at Cape Alonaki between Fira and Athinios shows an impressive section of Santorini's volcanic history. At sea level, the lowest visible rocks are lavas about 0.5 million years old. In the middle of the photo the Lower pumice series (Bu) is clearly recognizable; it is overlain by the Middle pumice (Bm) and, at the top right, the Minoan pumice series (Bo).

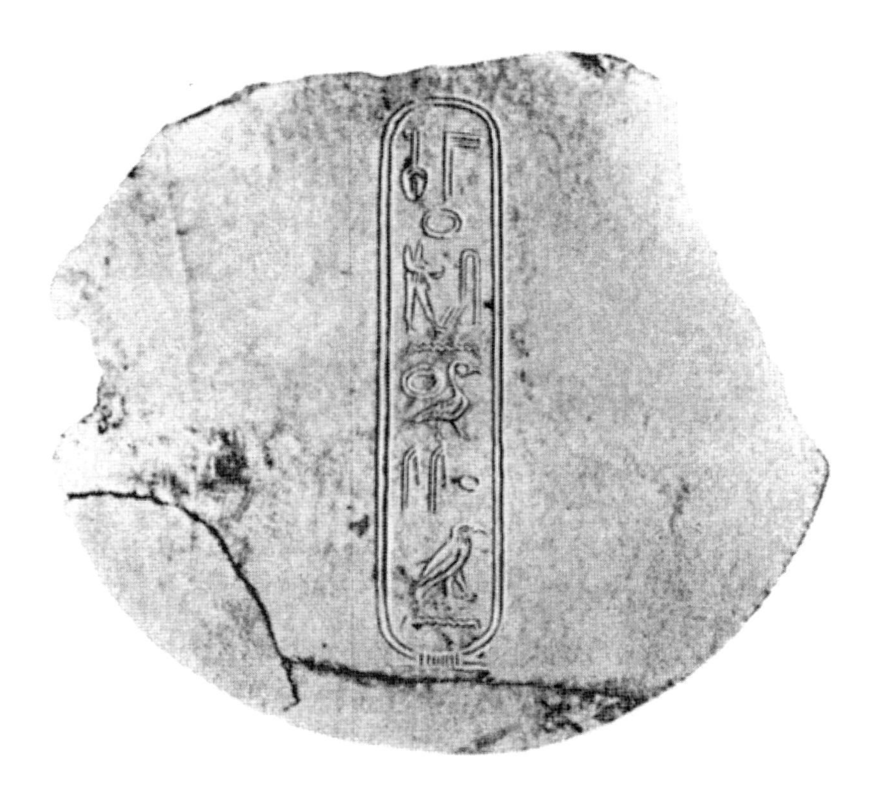

Figure 7.1 The cartouche with the name of King Khyan, who ruled Egypt during the period of Hyksos, is an important discovery that shows that the Minoans had connections of some kind with Egypt. It was excavated by the British archaeologist Sir Arthur Evans at Knossos on Crete. This finding has long been regarded as an important argument for an age of around 1500 BC for the Minoan eruption. From Evans (1928).

logical reference points to Egypt (Warren, 1990). Thus, we find ourselves wondering whether an inscription on the column at Thebes that reported flooding of the Nile delta during the reign of Ahmose I, who reigned from about 1539 to 1514 BC, can be related to the Minoan eruption (Davis, 1990).

A completely new perspective was opened by Minoan discoveries in Israel (Niemeier, 1990) and Egypt (Bietak, 1992). While the material found in Israel consists only of sparse remains of Minoan wall paintings (murals), the excavations of ancient Avaris in the eastern Nile delta consist of more than a thousand fragments that make possible a correlation with the Egyptian chronology based on the dynasties of pharaohs. The discoveries from Ezbet Helmi on the Nile delta originated from the so-called Hyksos period and the beginning of New Kingdom, which, until now, has been considered a dark chapter in the chronology of the pharaohs. The well-preserved fragments of frescoes with Minoan motifs of bull-leapers (Fig. 7.2) and labyrinths are among the most interesting discoveries of the past few decades. In the choice of motifs and elements of style they clearly show harmony with material found on Crete and Santorini. These extraordinary relicts have stimulated lively discussions and speculation among historians. Were the rulers of the Hyksos period on the throne of the pharaoh Minoans? Did their cultural influence reach Egypt through conquest or was there perhaps a Cretan princess on the throne of the Nile delta? The latter possibility was proposed in the reports of the leader of Austrian excavations, M. Bietak, in the *Frankfurter Allegemeine Zeitung* (27 July 1993). In a chronological sense, the Minoan artists that produced the wall frescoes on the Nile delta could have been survivors of the Minoan eruption who were forced to leave their homes and settle in Egypt. Certainly these new discoveries reopen the whole question

Figure 7.2 A fresco fragment from the Nile Delta shows three characteristic signs of the Minoan culture: a labyrinth, a bull, and a bull-leaper. The new findings at ancient Avaris date from the Hyksos period and the beginning of the New Kingdom when foreigners ruled Egypt. Drawn after a photograph in Bietak (1992).

of the age of the Minoan eruption. The absolute ages will probably give us an adequate answer to this question.

New attempts at dating

For several years a number of scientists have argued that a time span of 50 years may have separated the eruption and the collapse of the caldera. There were also experts who were of the opinion that the eruption could have been much later, in 1200 BC. The Dutch geologist van Bemmelen (1971) even believed that the 'Seven Egyptian Plagues' of the bible could be linked to the Minoan eruption. In particular, the observed darkening of the sky might have been a consequence of the ash of the Minoan eruption. The chronological relation seems very unlikely, however. As the work of Friedmann (1992) shows, the discussion of these topics is not yet settled.

Using methods of relative dating one can narrow down the dates of volcanic events of Santorini, but archaeologists are united in the belief that they must have occurred in the time range of 1800 BC to 1300 BC.

There have also been attempts to date material from Santorini by thermoluminescence methods. Vagn Mejdahl of the Risø Research Institute has dated potsherds from the excavations at Akrotiri using this method. He obtained an age of 3600 ± 200 years (Friedrich, 1987), which is in good agreement with results obtained by other methods.

Frost damage to trees of North America should be traceable to the Santorini eruption. Frost damage related to volcanic eruptions has been observed not only in America but in Europe as well. Baillie (1988) firmly established temporally consistent frost damage in the annual rings of oaks found in a bog in Ireland. At the third International Thera Congress, which was held on Thera in September 1989, Baillie made the following comment: 'Most likely, it is a question of the frost scars observed in the trees of Ireland and similar ones from America, but we do not know what caused this event.' This method has the serious disadvantage that climatic variations responsible for the frost damage can also be caused by other natural events, such as the shifting of ocean currents, known as El Niño, observed off the coast of Chile.

Dendrochronology, the counting and interpretation of annual growth rings of trees, or the wooden beams made from them and used in old buildings, has already provided many reliable archaeological and geological results. Attempts to apply dendrochronology to dating the Minoan eruption have so far failed, however, because at Akrotiri so far finds of wood are rare, and it is poorly preserved. Peter Wagner of the Botanical Museum of Copenhagen (Friedrich, 1990) described the anatomy of a tamarisk branch from the excavations at Akrotiri and also found several annual rings. But the material is insufficient for an age determination by dendrochronology, because it shows only about ten annual rings and at least 50 are required (Fig. 7.3). The material is quite suitable for radiocarbon dating, however, and Henrik Tauber of the radiocarbon laboratory at the National Museum in Copenhagen dated it at 3380 ± 60 radiocarbon years (K-4255).

The radiocarbon method is one of the most powerful dating techniques used in archaeology; it is also capable of measuring the age of very young geological material.

Samples from the Akrotiri excavations that have been dated in different laboratories have in many cases yielded unexpected results. Some samples could not be used or were of only marginal usefulness, because they were too small. Other samples of short-lived material such as fruit or seeds yielded an age that was about 150 to 200 years older than the archaeological estimates. An interesting phenomenon turns out to be the cause of this.

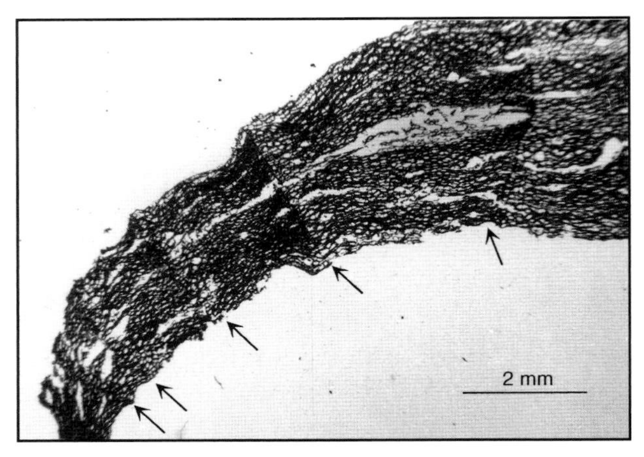

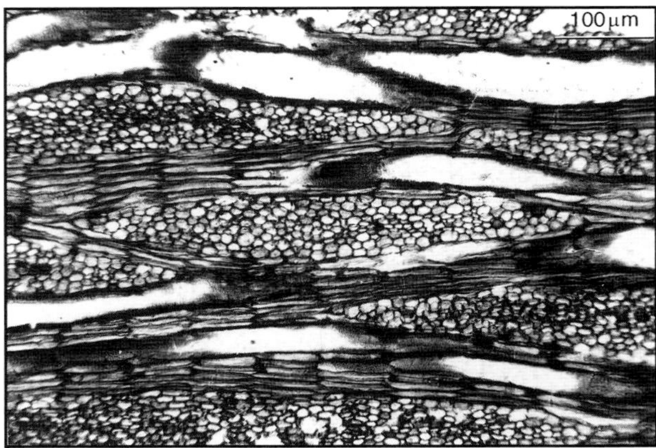

Figure 7.3 Annual growth rings are clearly visible in a twig of tamarisk from the Akrotiri excavation (arrows). Unfortunately, this piece of wood was not suitable for dating by dendrochronology, because it had too few rings. It could, however, be used for dating by the radiocarbon method; it gave an age of 3380 ± 60 radiocarbon years (Friedrich *et al.*, 1990).

It has been shown that the radiocarbon dating of modern plants growing in volcanic regions in the vicinity of carbon dioxide emissions gives completely useless ages. Such plants appeared to be more than a hundred years too old relative to those that grew in more distant settings. But it also shows that these effects are confined to plants growing near carbon dioxide emissions, so not all the samples of the surroundings are affected. The approximate values, which the majority of samples indicated, were used to designate an age and outliners were rejected.

At the radiocarbon laboratory in Copenhagen seeds of 'faba' beans (*Lathyrus clymenum*) and lentils (*Lens culinaris* Medik.), which were found in jars in the excavations at Akrotiri, were used for dating. The results agreed in general with those that had been obtained in different radiocarbon laboratories for short-lived material. This dating also agrees quite well with the values for a hearth which was found in the Kharageorgis quarry at Athinios on Thera under pumice from the Minoan eruption (Friedrich *et al.*, 1980b).

Meanwhile, there are some well-dated samples from the excavations at Akrotiri. All are of short-lived material. The mean values for radiocarbon ages (in years before 1950), measured at the following laboratories, are:

Philadelphia	3300
Heidelberg	3350
Oxford	3338
Copenhagen	3340

The mean value of the four measurements above is about 3340 radiocarbon years before 1950. When one takes these values and transfers them to the calibration curve of Pearson and Stuiver (1986), one gets a corresponding value in calendar years. For the example above, one obtains – by a remarkable coincidence – the value of 1645 BC, which is identical to the value obtained from ice cores. The ice core method has contributed remarkably precise data for the age of the Minoan eruption.

Our earth has several types of repositories in which history is preserved. For example, the wandering of the magnetic pole over many millions of years has left traces in the lavas on the ocean floor. The rocks of the lavas contain a record of their age in their isotopic content and their minerals have kept a record of the magnetic properties of the poles. There is even an archive where the climatic record of the past is preserved with quite suitable accuracy. It is recorded in oxygen isotopic ratios, $^{18}O/^{16}O$, in marine limestone. Thus one

Box 7.1

Two American scientists, LaMarche and Hirschboeck (1984), observed damage to the annual growth rings of the Californian bristle cone pine (*Pinus aristata*) that they attributed to frost damage. This tree has very thin annual rings that record thousands of years of slow growth in the White Mountains of California. Explosive volcanic eruptions that send large amounts of dust and acidic aerosols into the stratosphere can cause global climatic effects, as was seen from the frost damage caused by the catastrophic eruptions of Tambora in 1815 and Krakatau in 1883 (see Boxes 6.1 and 6.2). It is reasonable, therefore, that similar damage found in the growth rings corresponding to the period 1628 to 1626 BC was caused by an important volcanic event.

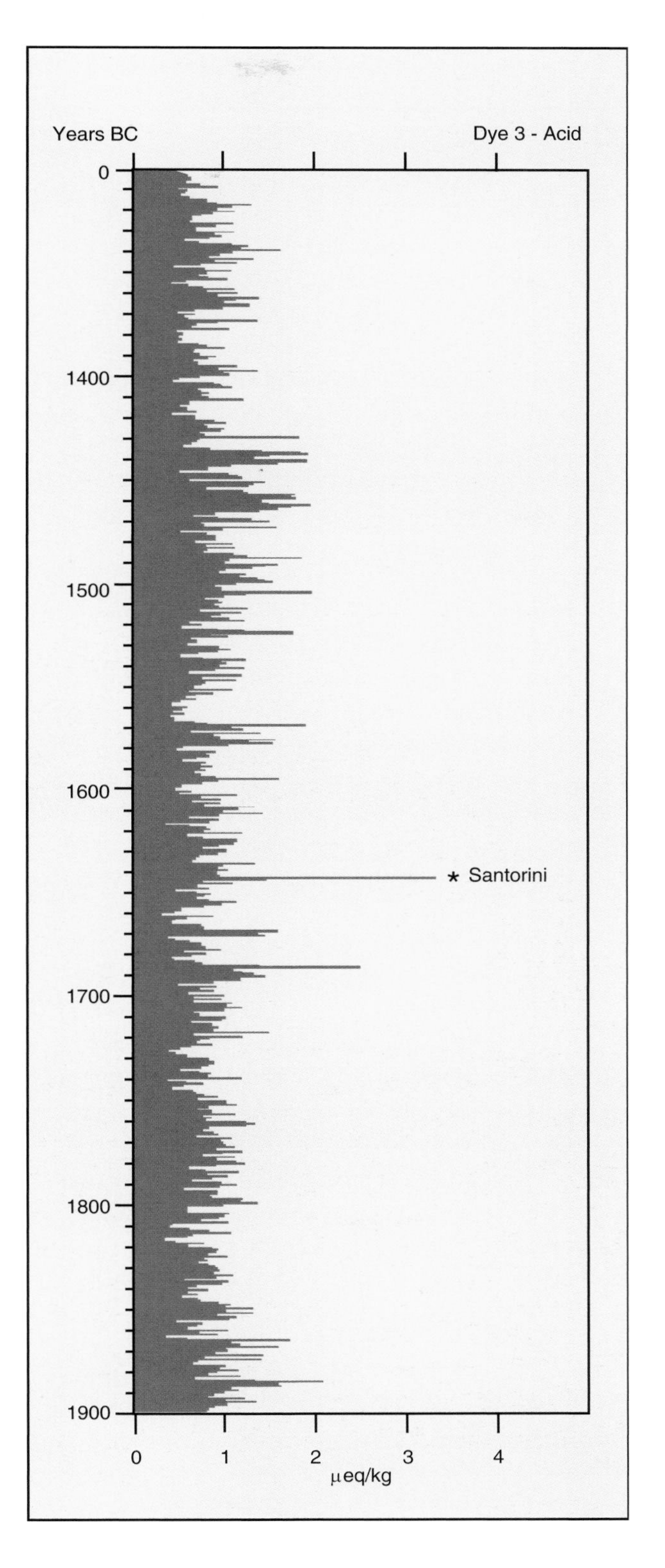

Figure 7.4 Each year's precipitation is stored in the inland ice of Greenland. In the ice core of 'Dye 3' in southern Greenland it is possible to count the annual layers by means of chemical and optical methods. Anomalous concentrations of sulfuric acid can be related to volcanic eruptions. The acidity peak resulting from the Minoan eruption is marked by an asterix. (After Hammer *et al.*, 1987.)

Box 7.2 The radiocarbon method

This method depends on the formation of radioactive ^{14}C by addition of additional neutrons to atoms of nitrogen when cosmic rays hit the atmosphere (Fig. B.7.1). It was previously thought that this process was going on at all times at a regular rate and that the amount of ^{14}C being formed was constant. It was later learned, however, that cosmic rays can become much weaker and that these changes affect the amount of ^{14}C produced and, in turn, the amount absorbed by living organisms. These changes have a great influence on the results of the dating: plants absorb radioactive ^{14}C in their tissue and it is transferred to the flesh of animals that feed on the plants. While the organism is living the amount of ^{14}C is kept at a constant level, but when the organism dies the amount is reduced at the rate of the half-life. The radioactive ^{14}C, which has a half-life of 5730 years, then decays to the stable nitrogen isotope ^{14}N. This means that after a period of 5730 years only half of the original ^{14}C amount remains.

This phenomenon provides a simple way of determining the age of organic material. If one compares the measured ^{14}C content of a dead organism to the original amount of ^{14}C one gets the age of the investigated object.

Anomalous amounts of ^{14}C have been found in the annual rings of very long-lived trees, such as the *Sequoia*, when selected rings of known ages are dated using the ^{14}C method. These variations were also found in living bristle cone pines (*Pinus aristata*) that had more than 4000 rings and in the remains of trees in the same area that died more than 4000 years ago (Fig. B.7.2). In this way it has been possible to establish a continuous record of annual rings reaching back about 9000 years, to which other material could be compared. Another dendrochronological record reaching back even more than 11 000 years was established in Germany using

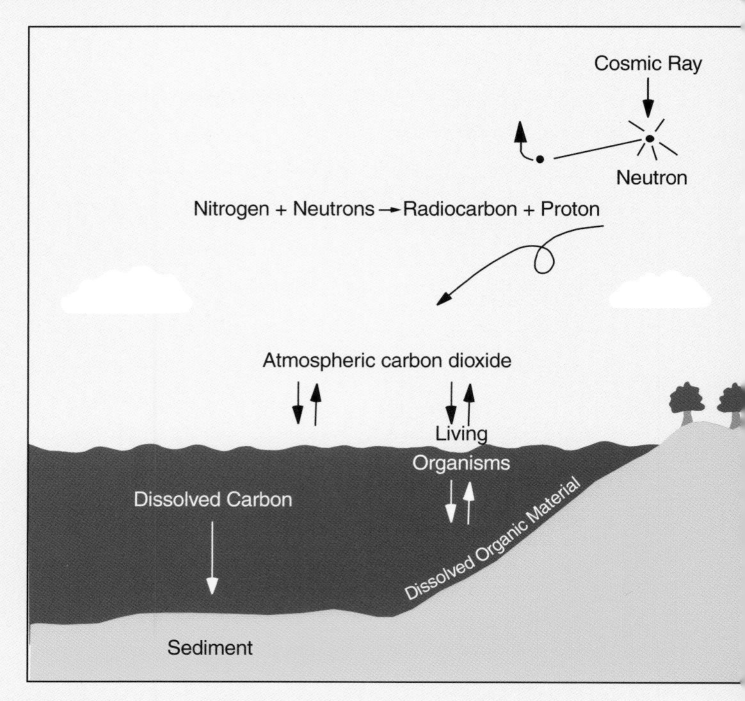

Figure B.7.1 Schematic diagram showing the origin of radiocarbon in the atmosphere (after Tauber, 1992).

Figure B.7.2 The exceptionally long-lived bristle cone pines of the White Mountains, California, provide a dendrochronological record to which radiocarbon ages can be calibrated. Some of these trees are about 4000 years old. LaMarche and Hirschboek (1984) found frost damage in the annual rings in the interval 1626–1628 BC that, according to their interpretation, were caused by climatic variations resulting from the Minoan eruption. From Löhr (1971).

Box 7.2 (*cont.*)
oaks and pines from the banks of Danube. With this dendrochronology one was able to fix the end of the last ice age at 10 970 'dendro years' before 1950 (Becker *et al.*, 1991). Continued research has made it possible to calibrate measurements beyond 11 000 years. Studies of the annual layers of corals in Hawaii may extend the calibration curve to about 30 000 years (Bard *et al.*, 1990).

Radiocarbon values that have been calibrated using this curve can be considered as 'calendar years', while uncalibrated data (i.e. all ages of more than 30 000 years) are called 'radiocarbon years'. In the time interval in which the Minoan eruption occurred, about 3600 years ago, an uncalibrated radiocarbon age can be 300 years older than the true calendar year. The possibility of a fine adjustment using dendrochronology gives new accuracy to the radiocarbon method.

Samples selected for dating should not have had an old initial age. The thick trunk of a tree, for instance, could have an age difference of several thousand years depending on whether one has measured the innermost or outermost annual rings. It is best to use short-lived organisms that were harvested during the same year. Thin branches with only a few rings can also be used.

A modern variant of the conventional ^{14}C method is the AMS method (Accelerator–Mass–Spectrometry) that can use samples of only a few milligrams. The results are of the same quality as by the conventional method. In cases where one has only small amounts of material, this has an advantage over the traditional methods.

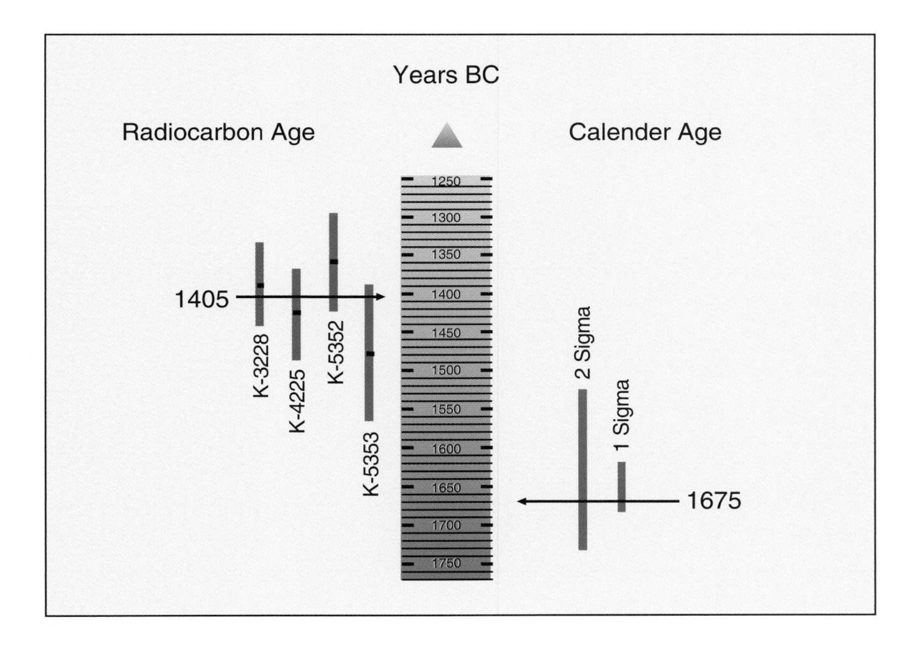

Figure 7.5 The radiocarbon dating by the Copenhagen laboratory of material from the Akrotiri excavation gives a mean value of 1405 years for the age of the Minoan eruption. Single grains of 'faba' (K-5352, K-5353, K-3228) and a thin twig of tamarisk (K-4225) were used as samples. Calibrating the *radiocarbon age* obtained by the curve of Pearson and Stuiver (1986) one gets an age in *calendar years* of 1675 BC. The sigma 1 band for statistical uncertainty indicates that there is a 68 percent chance that the real value is within the band. For sigma 2 it is 95 percent. (After Friedrich *et al.*, 1990.)

can, for example, determine from the lime in fossil belemnites the temperature of the water of the sea in which they lived many millions of years ago.

The presence of an acid signal in the Greenland ice showed that the eruption was strong enough for the eruption column to reach the stratosphere where the ejecta were carried over a large part of the Northern Hemisphere. Other methods predicted that the acid peak should lie in the time interval between 1800 and 1300 BC, and when it was found at 1645 ± 7 years BC, the date of the Santorini eruption was confirmed (Hammer *et al.*, 1987; Fig. 7.4). The ice-core date was also consistent with the mean value for calibrated radiocarbon dates (Figs. 7.5 and 7.6).

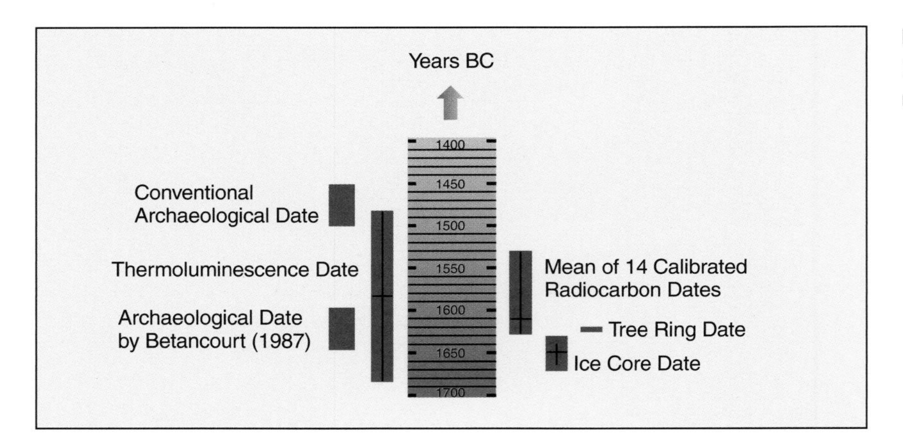

Figure 7.6 Dating results of the Minoan eruption obtained by different methods. (After Friedrich *et al.*, 1990.)

Box 7.3 Thermoluminescence dating

The thermoluminescence method of dating material that has previously been heated (such as ceramics) has been widely used in archaeology and, more recently, in geology. Energy from cosmic rays and other forms of ionizing radiation, such as alpha and beta particles or gamma rays from the natural radioactivity in material surrounding a buried artifact, accumulates in the natural minerals in pottery or other ceramic material.

This energy is stored in the form of electrons trapped in defects in the crystal lattice of grains of minerals such as quartz and feldspar. The longer the article is buried in the earth, the more electrons are stored. And since the rate of accumulation or 'dose rate' is usually constant, the number of stored electrons is directly proportional to the time that has elapsed since the last time the stored electrons were released by heating the minerals or exposing them to daylight. Heating ceramic material in a kiln 'resets the clock' and the minerals begin again to accumulate trapped electrons.

To measure the number of stored electrons the minerals are heated to about 500 degrees Celsius. As the temperature increases, the trapped electrons are released and give up their stored energy in the form of visible light known as thermoluminescence. The amount or 'dose' of this luminescence can be measured by a photomultiplier and the age of the artifact calculated by dividing the total dose by the dose rate.

This method has the great advantage that one does not depend on the chance preservation of organic material; the age of the sample is measured directly.

Box 7.4 Archaeomagnetism measurements

The principles of the archaeomagnetic method are based on two observations.

First, the direction of the magnetic field of the earth changes with time. The position of the magnetic pole wanders, and from time to time it switches polarity. Second, magnetic particles in geological or archaeological objects that have been at high temperatures take on the prevailing direction of the geomagnetic field when they cool below the 'Curie temperature'.

By measuring the orientation of the magnetic particles and comparing this with a temporal record of magnetic orientations based on well-dated samples, one can determine when the object acquired its magnetic orientation.

There is good reason, therefore, to accept this age obtained from analyses of 3200 meters of continuous core taken from the highest and thickest part of the Greenland ice sheet. The acid signal of Santorini has also been found in another borehole, referred to as 'Summit'. Thus, after many years of study and countless tests an accurate interpretation seems to have been achieved.

Box 7.5 The ice-core method

Some of the most interesting archives of information about the earth's history are the polar ice caps, where the precipitation of several thousand years is stored in stacks of ice layers. In Greenland, the ice reaches a thickness of about three kilometers, in which every year has left a characteristic ice layer. The layer of a single year no longer has its original thickness, since it has been compressed by the weight of overlying snow and ice, but it does contain chemical and optical characteristics specific to that year. We can count these annual layers just as we can count annual growth rings of trees, and in this way one can learn the age of a given layer. Counting of the ice layers is possible only through costly core drilling and an expensive laboratory is required to make the measurements. Several boreholes have been drilled into the Greenland ice sheet, but only three were deep: Camp Century, Dye 3, and Summit. The latter, which is situated in central Greenland, reached a depth of 3000 meters, while drilling close to the radar station Dye 3 produced a continuous ice core of 2300 meters (Fig. B 7.3).

The ice of the polar ice cap contains important information not only about the climate of the last 200 000 years but also about major volcanic eruptions (Fig. B.7.4). The record is not from the ash of the eruption but from sulfuric acid formed when sulfur-rich gases given off during an eruption reach the stratosphere and combine with water. These gases drift for great distances in the jet stream and are deposited in the normal precipitation of snow. The minute amounts of acid can be detected in an ice core because the acid increases the electrical conductivity of the ice. Two electrodes are moved in a zig-zag path along the length of the ice core while the electrical conductivity is recorded on a chart (Fig. B.7.5). When an acid signal is found with this instrument, this part of the core is analyzed by time-

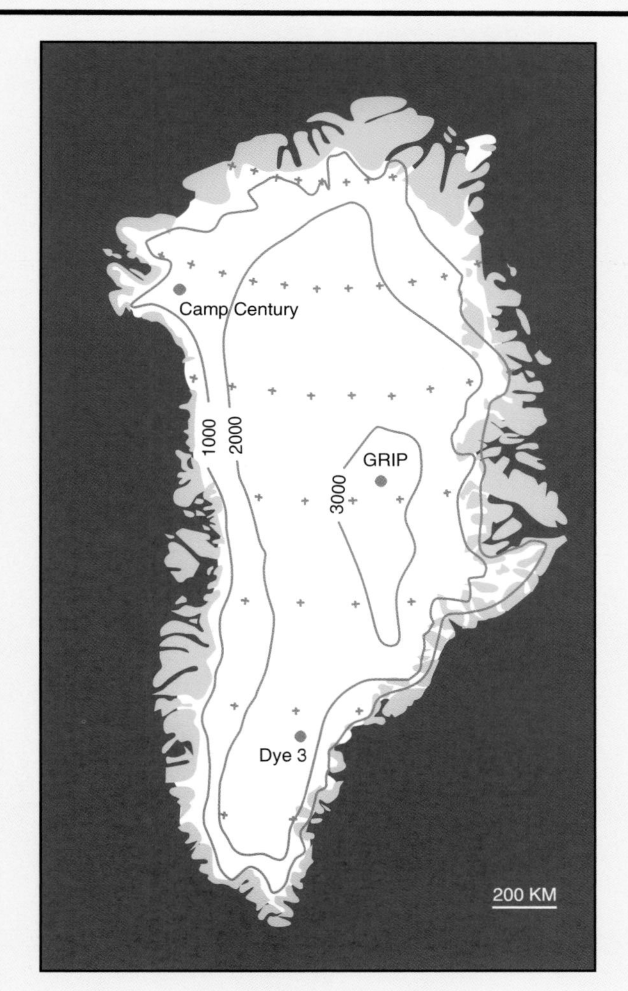

Figure B.7.3 The three ice core drilling sites, 'Camp Century', 'Dye 3', and 'GRIP Summit' (GRIP is the Greenland Ice Core Project) are all located in the Greenland ice sheet. The acid signal of the Minoan eruption was found at all three sites. The contour lines show the thickness of the ice in meters. Modified after Johnson *et al.* (1992).

consuming chemical methods. Each annual layer is characterized by its chemical composition and stored in a continuous record, so that the analysis can then be

Box 7.5 (cont.)

related to a date in the ice stratigraphy. This technique has been used by the geophysicist Claus Hammer, of the Geophysical Institute in Copenhagen, to detect the acidity signals of volcanic eruptions. Some of the precisely dated acid signals can be correlated with historic volcanic eruptions, such as those of Krakatau (1883) and Tambora (1815). The same acid signals are found in different ice cores, so the ice core method is considered very reliable. All eruptions are not equally well recorded, however. The strong eruption of Katmai in Alaska in 1912 was barely recognizable in the Dye 3 drilling. This was because the eruption occurred at a high northern latitude. Owing to the northerly direction of the stratospheric wind in the Northern Hemisphere, these acid signals were weak in the drill cores from southern parts of Greenland. Because individual acidity signals can be correlated with known eruptions, it is also possible to eliminate errors that might result from loss of core or other accidents. It is also possible to define the maximum error limit, but there is a complication. Since atmospheric nitrates also produce acid signals in an ice core, one must use chemical methods to distinguish peaks generated by melting of ice in very warm summers from sulfuric acid signals. This gives the ice cores an advantage over frost damage to trees, for one can be sure that the sulfuric acid in the ice core is produced by a volcanic eruption, while frost damage could have a variety of causes. The age obtained from an ice core is also more reliable because every annual layer is characterized by several factors. Enormous eruptions in the vicinity of the equator with the strength of that of Tambora are recorded in ice layers of both hemispheres. For example, the eruption of an unknown volcano in 1249 AD left traces of acidity in both Greenland and Antarctica (Hammer and Clausen, 1990), but this strong volcanic event is not recorded in the tree rings of the US.

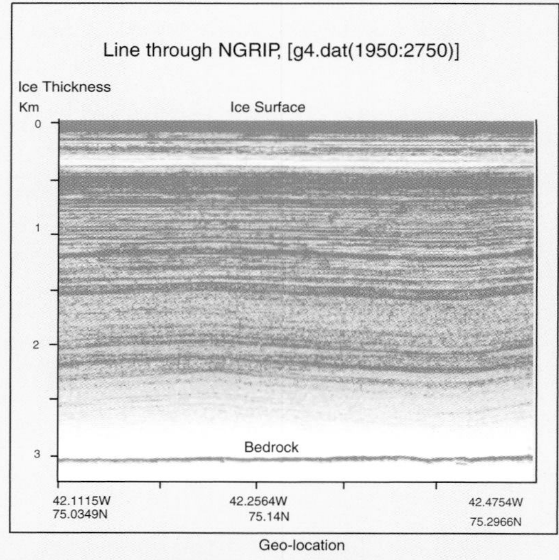

Figure B.7.4 The Greenland Ice Shield becomes transparent in radio-echo imagery. This made it possible to see through the entire 3-km- thick ice sheet and identify reflecting horizons that could be correlated with acidic volcanic layers found in the summit drill site in northern Greenland (NGRIP). (Dahl-Jensen *et al.* 1997.)

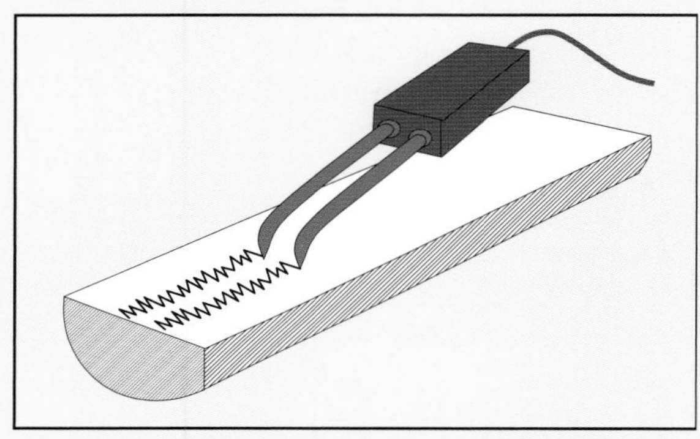

Figure B.7.5 In order to detect acidity signals, two electrodes are moved over the surface of a section cut along the length of the ice core. The electrodes are connected to an instrument that measures conductivity. In those places where the ice is anomalously acidic the conductivity increases. The acidity is caused by sulfuric acid produced by a volcanic eruption and taken up in water vapor that is precipitated as snow. The measurements require time-consuming processing of the ice in the laboratory. After Dansgaard and Hammer (1981).

It the summer of 1992 ice core drilling at Summit (Fig. B.7.6) was carried out successfully by eight European countries cooperating in the Greenland Ice Core Project (GRIP). A team of as many as thirty scientists spent four summers on the Greenland ice. On July 12 the bottom of the inland ice was reached at a depth of 3028.6 meters. At this point the ice has an estimated age of 200 000 years. This continuous ice core represents a database with information covering the last ice age and interglacial

Box 7.5 (*cont.*)

periods. As the longest continuous ice core that has been drilled in the Northern Hemisphere, it is an ideal source of information for this time interval, because there were no disturbing melting phases of the kind found in the ice core from Dye 3 in southern Greenland. The first scientific results were presented in the report of Johnsen *et al.* (1992), who reported the results of the study until the year 1991. Of special interest with regard to Santorini were the acidity signals that had already been observed in Century and Dye 3. They were found at a depth of 736.45 meters, which represents an age of 1636 ± 7 years BC (Clausen *et al.* 1997). As a result, there are now two very valid dates for the same event, one from Dye 3 and the other from the GRIP. The mean of the two (1644 BC and 1636 BC) gives a value of 1640 BC. It should be mentioned that in 1994 Zielinski *et al.* published an ice core date of 1628 BC, identical to that obtained from the frost damage on trees, but scientists working with ice cores (Clausen *et al.*, 1997) consider this value less reliable. At this time, therefore, 1640 BC is considered the best date for the Minoan eruption.

Figure B.7.6 Recovering an ice core from the drilling at Summit in central Greenland, where the acidity signal of the Minoan eruption was detected. (Photos W. Dansgaard, July 1993.)

Part 3

The volcano releases its secret

8

A Bronze Age Pompeii

Quite unexpectedly, a 'Bronze Age Pompeii' was discovered in 1866 under a blanket of pumice on Santorini. For more than 3600 years, its secret had been concealed, even though vague memories of the volcanic catastrophe had been preserved in the legends of Greece and Egypt.

The first discoveries on Therasia

The first traces of a Bronze Age settlement on Santorini were found in the 1860s on the island of Therasia. At that time, there were large quarries in the northern and southern parts of the island where pumice was excavated and shipped to the Suez Canal for construction of Port Said. The 'Santorini-earth' or 'pozzolana', as the pumice was called, produced, when mixed with lime, excellent cement that hardened well under water. For centuries, cement of this kind had been used for piers throughout the Mediterranean. Pumice was easily available in abundance at many places on Santorini and it quickly became an important resource for use by the Suez Canal Company. There were also large quarries on Thera north of Akrotiri and on the caldera rim at Athinios, as well as on Oia, all of which are now shut down. However, the scars on the landscape are still easily recognizable today.

In one of these quarries a sensational discovery was made that suddenly threw light on the historical importance of Santorini. On the east side of Therasia, between Cape Tripiti and Cape Kimina opposite the island of Aspronisi, workers in the Alafousos quarry uncovered the remains of walls beneath the pumice (Fig. 8.1).

The significance of this discovery was not appreciated until it became known to scientists who came to Santorini in 1866 to observe the eruption on Nea Kameni. Among the investigators was the chemist

Figure 8.0 The site that Robert Zahn had excavated at Kamaras in the valley of Potamos is only a hundred meters or so from the present excavations at Akrotiri. At the turn of the century he worked close to the red ignimbrite platform (middle right of the photo), where he found Bronze Age remains that are now stored in the archaeological museum in Thera.

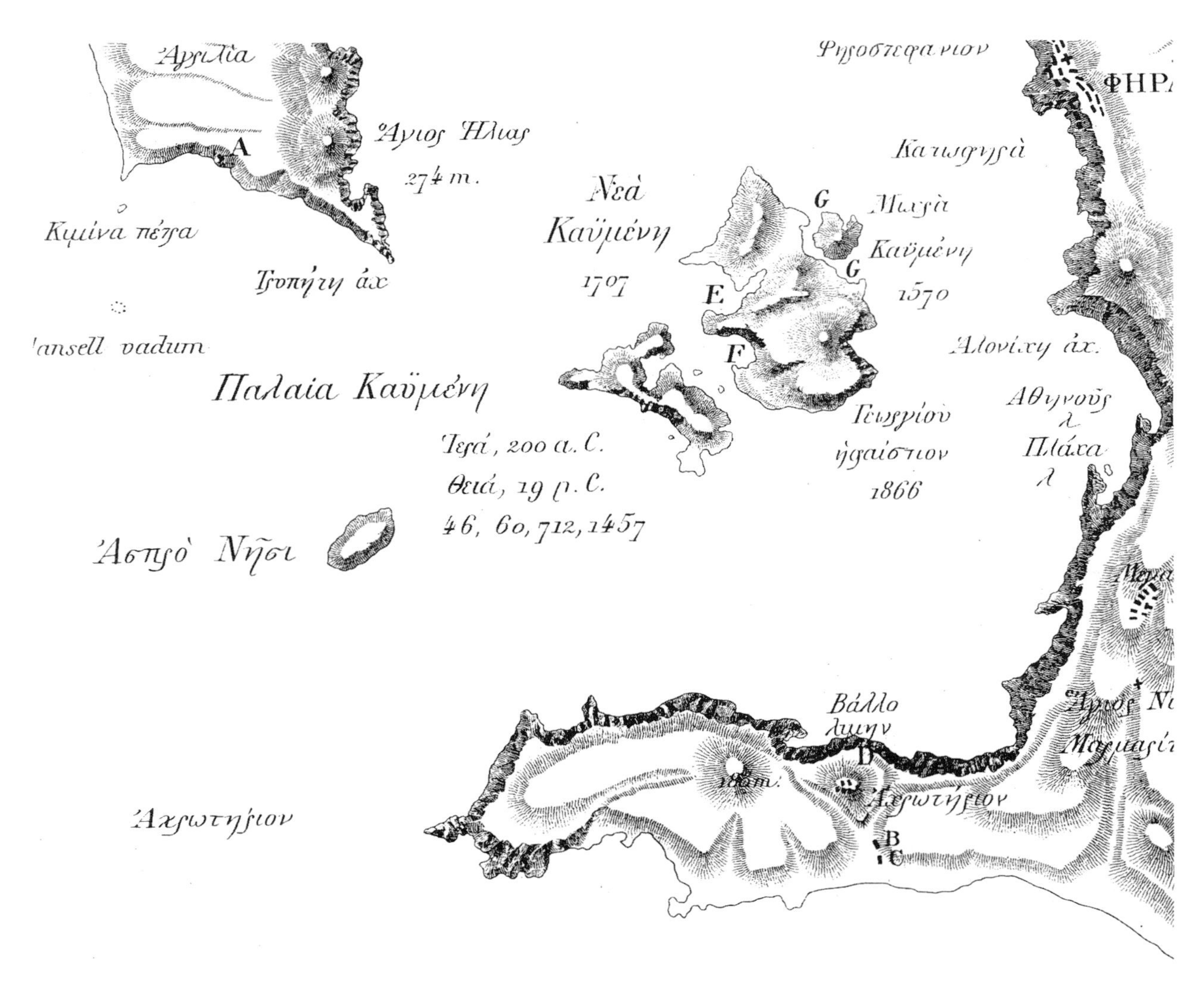

Figure 8.1 Locations of ruins from the Bronze Age on Thera and Therasia are seen in an enlarged section in the map by the French archaeologist Mamet (1874). On Therasia they are located in the Alafousos quarry (A). On Thera they are at Akrotiri (B and C) and at Balos (D).

Christomanos of the University of Athens. When, by chance, he visited the quarry he immediately realized that the ruined walls were older than the pumice that covered them. It was clear to him that it could not be a question of later tombs dug into the pumice, as was then thought.

The owner of the quarry, Alafousos, and a doctor from Thera, Nomicos, took over the excavations of the discovery. They uncovered a house with several rooms in which they found ceramics and tools.

Among the other scientists who came to Santorini was the French volcanologist, Ferdinand Fouqué. When he saw the ruined walls on Therasia, in 1867, he too was impressed with this extraordinary archaeological discovery. A new Pompeii had been uncovered. At the time of Fouqué's visit, the jaw bone and pelvis of a man had recently been found in the ruins, giving the impression that humans had been overwhelmed by the eruption, as they were at Pompeii. Fouqué, together with other specialists, pursued the excavations, and reported the discoveries in a short paper entitled 'Une Pompëi antéhistorique' (1869). More details were provided in Fouqué's later, more comprehensive work, *Santorin et ses Eruptions* (1879), in which he reproduced drawings of some of the discoveries on Santorini. This excellent book is now available in an English version translated by A.R. McBirney (1998). At Fouqué's time it was not yet known that the cultural remains buried under pumice ranged in age from the Stone Age up till the Minoan period. The term 'Minoan', which at that time was unknown, was first used by Sir Arthur Evans at the end of the nineteenth century with reference to the palace of Knossos he uncovered on Crete.

After examining the position of one of the buildings, which was on an extensive base of lava, and the orientation of windows, which were looking out in the direction of the bank of pumice, Fouqué concurred with the perceptive view of Christamanos that it was not a tomb but was rather a dwelling (Fig. 8.2). The building style was quite unlike that seen on Santorini today. The walls consisted of lava and soil. In places, branches of trees were inserted into the masonry. From the form of these branches and the bark that was still recognizable on them Fouqué concluded that they must have been the wood of olive trees.

The wood was already so badly decomposed that it disintegrated into dust at the slightest touch. Wood from the later Akrotiri excavation was also in a similarly bad state of preservation. No metallic objects were found, only ceramics, a flint sickle, and the point of a spear eight centimeters long.

Today, the excavation sites are more accurately located. One can still, of course, examine the former quarries, but the ruins found in the first discoveries have long since deteriorated. Later investigators, such as Mavor (1969) and Aston and Hardy (1990), found only fragments of pottery in these areas. Apparently, as Fouqué sadly anticipated, valuable archaeological treasures at these sites had been destroyed. In this respect, the situation on Santorini 100 years later had not changed.

The archaeologist Christina Televandou (1989) remarked, in regard to the findings in the Mavromatis quarry near Akrotiri, that the quarrying activity in this region is a 'real calamity for the prehistoric culture and the history of the Cyclades' and should be terminated once and for all. Today all pumice quarries are shut down, and the Mavromatis quarry is now protected.

The discoveries on Thera

Fouqué found more ruins in an erosion channel near the village of Akrotiri on Thera. Here there were the remains of walls, which extended into the pumice and resembled those of Therasia. Fouqué was not able to excavate them, however, because he could not obtain permission from the owner.

Fouque's exploration was later carried out by the systematic excavations of Gorceix and Mamet of the French School of Archaeology in Athens (1870). Mamet published a note on these investigations in 1874 in his doctoral thesis under the title *De insula Thera* [The Island of Thera]. Even though his work contained only a brief sketch of the excavations, it was clear that he was dealing with the same locality at which the excavations at Akrotiri are now being carried out, 120 years later.

Fouqué left a fascinating account of the discoveries at Akrotiri. He described the excavations of a room and a Bronze Age fresco, apparently the first revealed since the eruption (Fig. 8.2B). He also described the colors: bluish red, yellow, dark brown and bright blue, which quickly faded on exposure to the atmosphere.

Later, Mamet and Gorceix continued their excavations at Akrotiri, where they unearthed a small house. In addition, they found a house north of Akrotiri in the vicinity of the bay of Balos. It was exactly on the rim of the caldera. Among the artifacts they recovered were some jars in which they found the remains of barley, lentils, and chickpeas. They also reported finding a number of ropes, many of which were braided on pieces of wood which, if disturbed, disintegrated into dust.

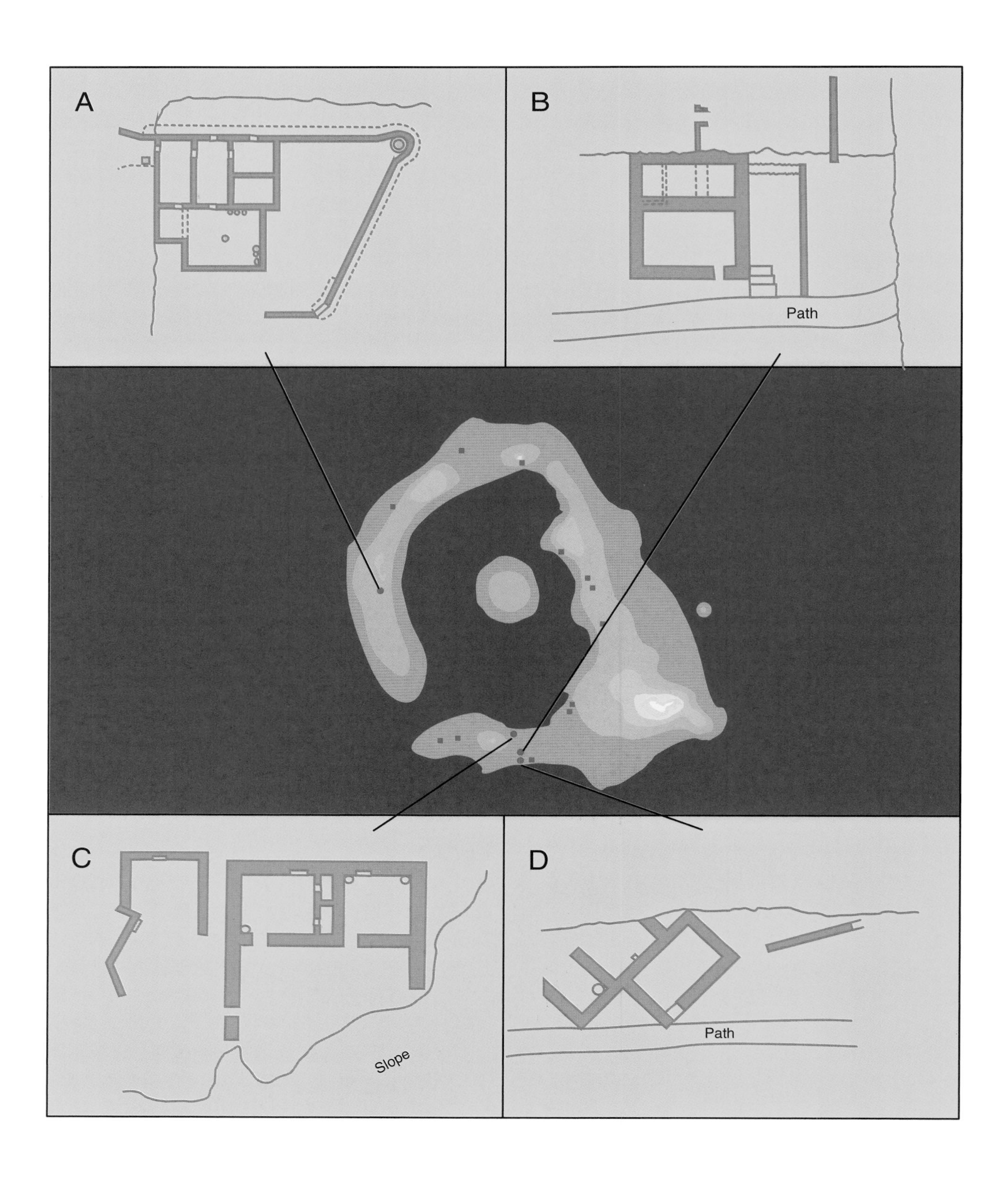
A
B
Path
C
Slope
D
Path

Figure 8.2 These diagrams of the ruins at the four Bronze Age sites shown in Fig. 8.1 were drawn by Mamet (1874) and reproduced by Fouqué in 1879. Their locations are shown on a more recent map showing the form of Santorini as it was during the Bronze Age. The distance between the contours for elevation is 100 meters. A: a house in the Alafousos quarry on Therasia. B: ruins that were excavated by Mamet and Gorceix in an erosional channel near Akrotiri. It was at this site that the first frescoes were uncovered. C: a house on the caldera rim at Balos. D: a house in the erosional channel at Akrotiri. The red squares on the large map indicate Bronze Age site discovered in the twentieth century.

It was already clear to scientists who examined these excavations that the prehistoric inhabitants had a highly developed culture. This was evident from the wall frescoes, the painted ceramics, and the worked metals, such as copper and lead. Agriculture and trade with the neighboring region had reached sophisticated levels, and they even had trade relations with North Africa. From the findings of his archaeological colleagues, the French scholar Louis Figuier concluded in 1872 that Plato's account of Atlantis could have been based on Santorini. I shall discuss this possibility in greater detail later.

The excavations on Santorini were resumed in 1899, this time by the German archaeologist Zahn. He did not dig at the same places at Akrotiri as his French predecessors, but in an erosion channel at Kamares only about 100 meters to the north. This small valley was named 'Potamos', because in the rainy season it contained a stream (Greek – potamos) (Fig. 8.0). It was parallel to the erosion channel in which the excavations at Akrotiri are being conducted. The results of Zahn's excavations were briefly mentioned in the monumental work of Hiller von Gaertringen (1904), who dug around the end of the century at the site of Ancient Thera at Mesa Vouno. Åberg (1933) depicted some of the ceramic objects from Zahn's diggings, in his work on the chronology of the Bronze Age. The site of these discoveries at Kamaras was later filled in and is no longer visible, as Spyridon Marinatos found when he began his excavations at Akrotiri in 1969. Christos G. Doumas (1983) reported that two elderly citizens of Santorini living at that time could remember the excavations at Kamaras. Moreover, it could also be located in old photographs. In these photos one also sees that the Bronze Age settlement had stood directly on the red 'Cape Riva ignimbrite', the same stone used to construct some of the buildings at Akrotiri.

A few years ago, the remains of frescoes were unearthed in the course of digging a cistern in the vicinity of Zahn's excavation site. These and other

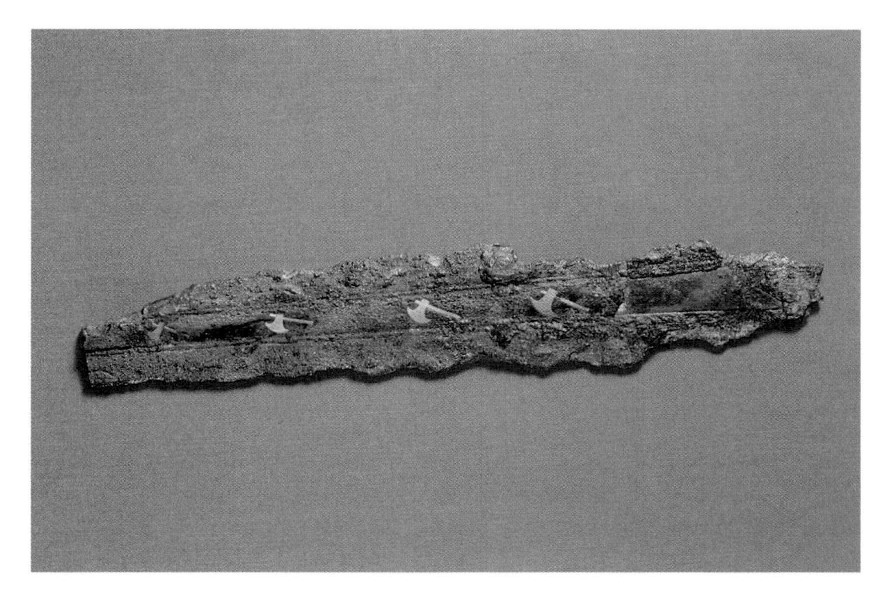

Figure 8.3 This bronze dagger with inlaid axes of gold was found near Kamaras at Potamos on Thera. It is kept today in the National Museum in Copenhagen (3167). The form and workmanship of the golden axes do not resemble those of Minoans but are similar to those seen on a similar object from Mycenae on which hunting scenes are depicted. For this reason it is thought that the dagger could be from Mycenaean times. Pottery shards from this period are known from another site on Thera. A fresco from the Akrotiri excavations shows a sea battle that is thought to have been fought between the people of Thera and the Mycenaeans.

Figure 8.4 A painted storage vessel (pithos) from a Bronze Age house at Akrotiri that was excavated in the nineteenth century. It is about one meter high. (After Fouqué, 1879.)

observations indicated that the Bronze Age settlement at Akrotiri was more extensive than at first appeared and might even have extended as far as the modern village of Akrotiri. The leader of the excavations today, Christos Doumas, postulates that the total area of the former settlement was about 200 000 square meters (Doumas, 1983). The discovery at Potamos, at around the beginning of the twentieth century, of a bronze dagger with inlaid axes of gold opens the possibility of further discoveries of importance in this area (Fig. 8.3).

When Spyridon Marinatos attempted to locate the excavations of his French predecessors of the nineteenth century, in 1967, 100 years had passed since the first Bronze Age ruins had been found on Therasia. In this time interval only Zahn had explored at Kamaras. The French archaeologist Renaudin (1922), who had studied artifacts from Akrotiri and Therasia, was able to classify it as late Minoan 1A (Figs. 8.4 and 8.5). Marinatos had already completed the first successful diggings in an erosion channel at Akrotiri and had renewed the 'Bronze Age Pompeii' discoveries. From the beginning, his extensive excavations at Akrotiri quickly led to almost daily sensational discoveries, including multi-stored houses embellished with admirable frescoes and painted ceramics (Fig. 8.6). Clearly, a very significant site had been uncovered, perhaps the most important in Greece of the century.

Years BC	Dynasties in Egypt	Crete	Greek Mainland	Santorini
	19th Dynasty			Late Minoan III
1300				
1400		Late Minoan	Late Helladic	Gap
	Amarna			
1500	18th Dynasty			Late Minoan IA
1600	Hyksos			
				Eruption
1700	13th and 14th Dynasty			
1800			Middle Helladic	Middle Cycladic
	12th Dynasty	Middle Minoan		
1900				
2000				
2100	11th Dynasty			Early Cycladic III
2200	I. Intermediate Period			
2300		Early Minoan	Early Helladic	
2400	6th Dynasty			
2500	5th Dynasty			Early Cycladic II
2600	4th Dynasty			?

Figure 8.5 Chronology of the Bronze Age in Egypt and Greece (partly after Doumas, 1983).

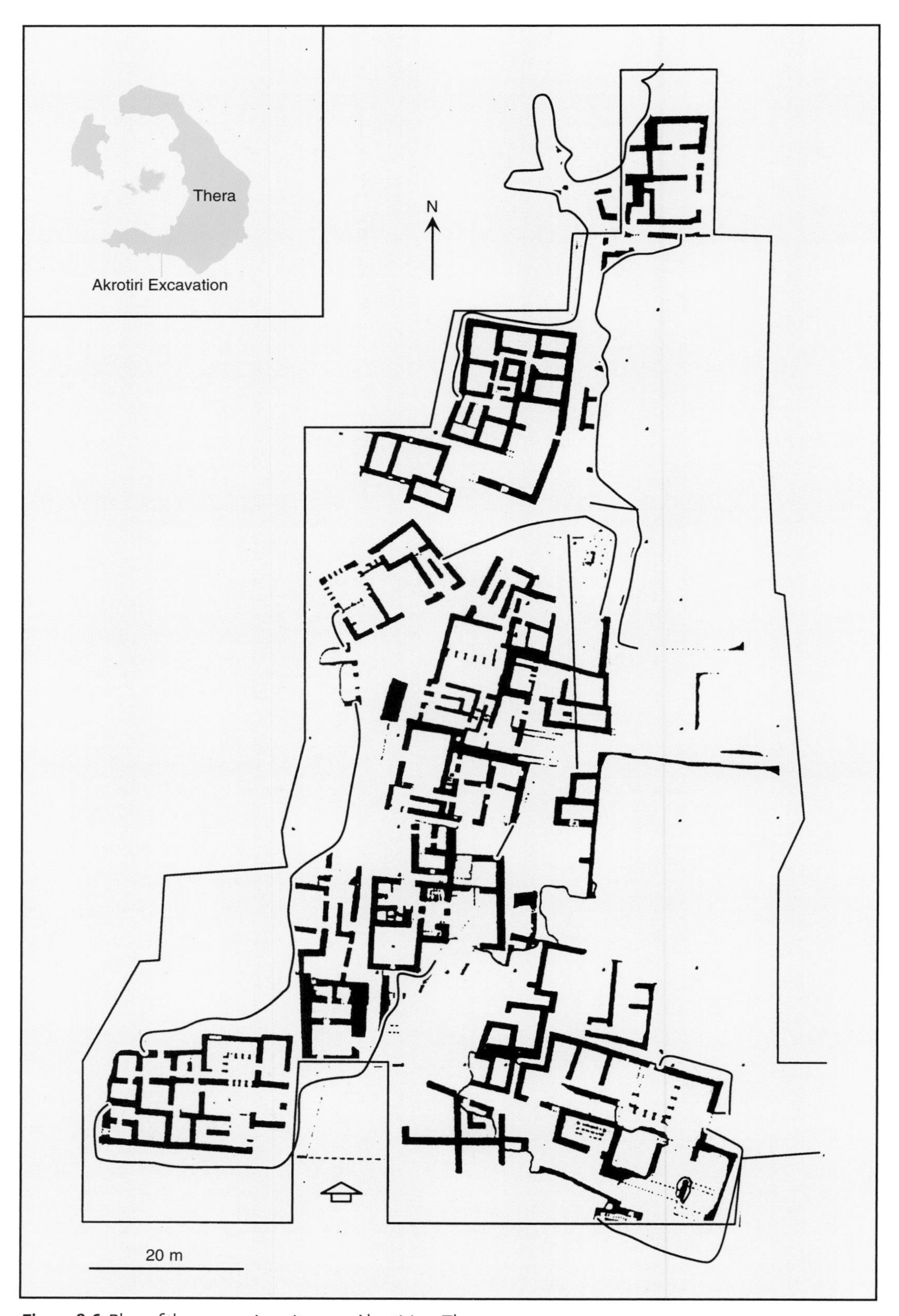

Figure 8.6 Plan of the excavation site near Akrotiri on Thera.

9

What was living on the island during the Bronze Age?

From their painted ceramics, their magnificent wall paintings, and the contents of their storerooms we can obtain a glimpse of life of the inhabitants of Akrotiri before the Minoan eruption. The discoveries reveal the nature of the flora and fauna at that time, as well as the commerce carried on with other parts of the Mediterranean region.

Plants and animals

We would know very little about the Bronze Age plant and animal life of Santorini had the people of Thera not left us two important sources of information: the food found in storerooms at Akrotiri and the animals and plants portrayed on frescoes, on jars, and even in sculptured figures.

Certainly, one must exercise caution in interpreting the drawings of life forms found on Santorini. Not all animals or plants depicted there were necessarily on the island at that time. Some are immediately recognizable as color patterns or symbols, such as the winged griffin; others, such as the blue monkeys, Sömmering gazelles, and ostrich eggs, are clearly imported from North Africa or the Nile delta. The potters had skillfully made one of the latter into a vase (Fig. 9.0). Also, the drawing of a river landscape in a fresco in the west house (room 5) at the excavation at Akrotiri is more suggestive of a landscape on the Nile delta than of the Bronze Age island. One sees on this particular fresco palms on the riverbanks, a monkey, flying ducks, and a running predatory cat.

The flora

On the islands of Thera and Therasia there are a few places where the older land surface of the island can still be seen. These are mostly areas where the pumice cover of the Minoan eruption has been stripped away by erosion or by quarrying during the last century.

Figure 9.0 An imported ostrich egg that has been converted into a vase indicates that the Thereans had trade contacts with North Africa. From the National Museum, Athens, courtesy of Professor Doumas.

One can search these early surfaces for traces of plants or animals and, as a rule, find nothing. This could be due to two causes. Either the eruption removed all traces or at the time there was simply no vegetation. If we accept the latter explanation, we may conclude that the few plants living on the island grew mainly in streambeds and other depressions sheltered from the wind, as in other volcanic regions where strong winds are prevalent. The former streambeds are today filled with pumice and not easily accessible. Only where the pumice has been excavated, as in the quarry of Fira or in the excavations at Akrotiri, does one find a few remnants of the earlier vegetation. But there, too, the plants are not preserved or have left only faint traces. For example, in the timbered framework of walls and in construction where branches or beams served as supports or braces, today there are only layers of pumice or cavities. Most of the original organic material is long since gone. The same holds for wooden furniture, such as beds, tables, and chairs, that the inhabitants left behind. But one can fill the cavities with plaster and make castings of the form of the wooden object that show which types of wood were employed at that time (Fig. 9.1). Presumably, where long straight beams were needed for construction, the wood of cypress or other conifers was used, but there were probably no such trees on the island at that time.

In this connection, the splendid ships of the people of Thera, such as the one seen in the so-called 'ship fresco' of Akrotiri (see Chapter 10, Fig. 10.26), should not be overlooked. Little of the wood needed to build ships of this kind could be found on Santorini; it had to be imported. Local sources of wood, such as olive and tamarisk, could be used to build houses, especially for the framework, and, of course, it could also serve as firewood. Since there is not the slightest trace of a forest on what remains of the earlier island, one must conclude that good wood for construction of houses and fine articles was imported from Crete, Naxos, the Greek mainland or Anatolia. Only tamarisk wood occurs in the excavations at Akrotiri (Chapter 7, Fig. 7.3) and olive wood in those of Therasia. Of course, the identification of this wood rests solely on the fragile remains of the branches and bark (Fouqué, 1879), but the discovery of an olive press in the excavations at Akrotiri confirms the existence of this tree on the Bronze Age island.

It would be helpful to compile a catalogue of the plants from the Akrotiri excavations, but although I have been trying to do this for several years my list is still incomplete, because the processing of the material found there is still in progress. Important new information continues to appear. For example, the Greek archaeologist Anayia Sarpaki (1990) carried out an important study of the remains of fruit and seeds of cultivated and wild plants found in storage jars at Akrotiri. Her work casts new light on the Bronze Age agriculture and vegetation of the island. In the excavations at Akrotiri, one finds in the walls of the houses impressions of Italian reeds (*Arundo donax*) with their characteristic longitudinal fluting. Evidently, at that time the long reeds of this plant were used to provide shade or to fill spaces over shelves and in ceilings. Bamboo-like plants of this kind are still used today on Santorini to construct awnings.

Baskets and eel-baskets were also used at that time. The slender and very flexible branches of monks pepper tree (or chaste tree) (*Vitex Agnus castus*) may have been used, as they still are today on Santorini for making baskets.

There is no question that the Bronze Age inhabitants of Thera were familiar with wine grapes (*Vitis vinifera* L.). Containers were decorated with clusters of grapes and it is reasonable to suppose that wine was also kept in the storage jars of Akrotiri. Moreover, very clear impressions of grape leaves are found in the Bronze Age settlements of Chania on Crete.

A small storage container from the Akrotiri excavations contained gray soil in which numerous almond-shaped cavities were distributed throughout. Plaster castings of these cavities clearly confirmed that the fruit of the almond tree (*Prunus amygdalus* L.) formed them (Friedrich, 1980). In some of the holes the

Figure 9.1 Nearly all objects in the Akrotiri excavations that were made of wood have long since decayed, but when they were surrounded by pumice their form was left as a mould that can be filled with fluid plaster, as is done to reconstruct human bodies at Pompeii. The shapes of beds and other furniture have been reconstructed in this way. The photo shows a profile in the middle of which one sees two vertical cavities that were left from a doorframe. In the area of the door opening one can recognize the pumice of Bo_1. Cavities were also left from the load-bearing beams used as supports for the next floor.

remains of almond shells could still be found, and in one of the cavities I even found the brown skin of a kernel (Fig. 9.2).

The conditions in which this material was found mean that the almonds were placed in the container mixed with soil, probably in order to keep them fresh and possibly to keep them from animals. We know from descriptions from Roman times that walnuts were stored in a similar way in sand (Lenz, 1859).

Fig trees (*Ficus carica* L.) must have grown on the island at that time. The remains of figs have been found in a clay jar in the Akrotiri excavations. When these fig remains were examined more closely in thin section, it was possible to make out their small seeds (Friedrich, 1980; Fig. 9.3). When the main, dark part of the fig was separated from the earthy material and the cavity filled with plaster, the original form of the figs could be reconstructed. The fruit was much

Figure 9.2 A vessel filled with earth contains numerous almond-shaped cavities. Filling these cavities with fluid plaster one could restore the original shapes of the almonds. Some of these cavities even contained the thin, brown skin of the kernel (below).

smaller than that growing on Santorini today. Figs are known from many Bronze Age settlements, such as Tyrins on the Greek mainland (Willerding, 1973), where they were just as small as those on Santorini.

Lentils (*Lens culinaris* Medik.) and vetch (*Lathyrus clymenum* L.) were also found in the storage jars of Akrotiri. The lentils resemble those of other Bronze Age localities in that they are very small. In cross section they measure less than 3 mm. Lentils were obviously an important component of the diet, as they are known to have been in other cultures of the Mediterranean region. One of the stories handed down to us in the Old Testament involves a dispute between Esau and Jacob over a red pottage of lentils for which Esau sold his birthright (*Genesis* 25:30–34). The lentils were probably the red Egyptian variety. In

Figure 9.3 The dark mass of an earth-filled vessel could be studied in detail: after hardening it with synthetic resin, the surface was polished and the tiny seeds of the figs were clearly visible. A plaster cast of a fig is shown below.

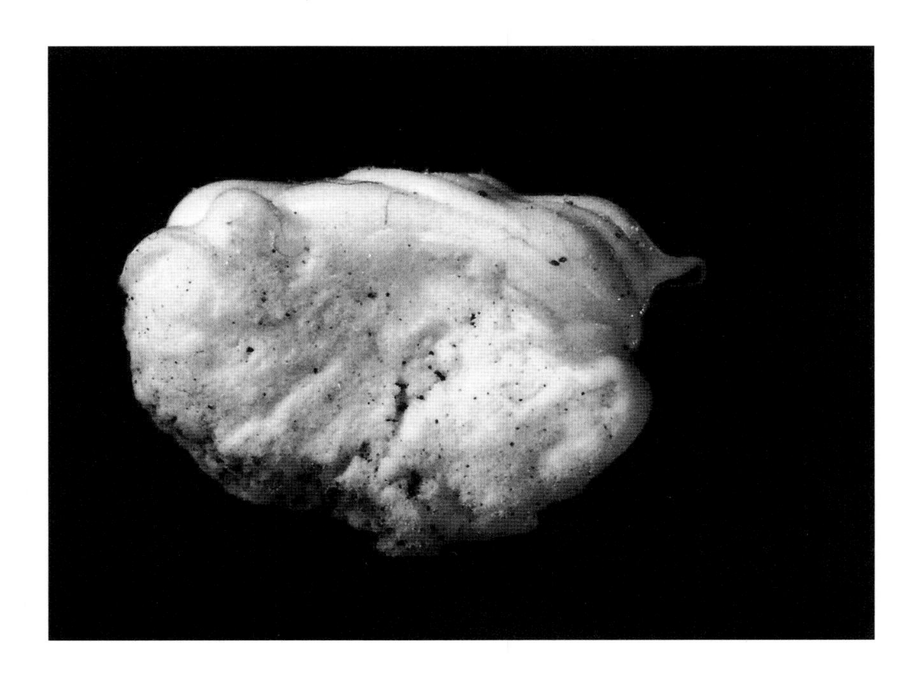

many other accounts from classical antiquity we read of the use of the lentils in the daily diet. Mixed with barley, for example, they were used to make bread. At that time, lentils were the principal food of poor people. We learn this from, among others, Aristophanes (about 455–387 BC) the poet of Attica. He said of the citizens of Athens: 'Times are now prosperous here, and the lentils are no longer scarce, but at that time, not only the poor but all the people ate them if they were able to obtain them.' Also Dioscorides, a Greek physician and pharmacologist, remarked in his work *De materia medica* (about 69 AD) that a diet consisting largely of lentils is not good for the eyes and can cause a bloated stomach.

The small *Lathyrus* seed was found in greater amounts in the storage jars of Akrotiri. Today these seeds are referred to on Santorini as 'faba', and they are still an important staple of the diet there. Such seeds are found in other excavation sites in the Mediterranean region. For a long time they were incorrectly taken for *Vicia faba* L. (var. *minor*), but the Greek archaeologist Anaya Sarpaki (1990) has recently shown that they were in fact seeds of the plant *Lathyrus clymenum*.

In the Bronze Age period of Akrotiri, faba seeds seem to have been stored in a manner similar to that which is still customary on Santorini today. They are dried in the sun and ground with a millstone

Figure 9.4 Even today, the people on Santorini use a stone mill to break 'faba' seeds almost exactly as their ancestors did 3600 years ago (see Fig. 9.5). Faba (*Lathyrus clymenum* L.) is still grown in the area of Akrotiri just as it was in the Bronze Age. Faba is a very popular food that is especially important for religious festivals. The people here are obviously following a very old tradition. In the background of the photo one can see the excavation site of Akrotiri.

Figure 9.5 A working place in the Akrotiri excavation shows a jar, a bench, and, on the floor, a stone mill.

(Figs. 9.4 and 9.5). After this, the chips are placed in a sack, which is stored in an earthenware jar (Fig. 9.6). The jar is closed with a wooden cover held down with a stone weight (Friedrich *et al.*, 1990). The grist is separated from the husks, and the seeds are divided into two or more parts. In this way the seeds dry more quickly and the larvae of beetles in them die. Sieving and winnowing removes the impurities. This method is not entirely successful, however, for I have identified the larvae of beetles such as bruchids on seeds from the excavation.

A small jar uncovered at Akrotiri contained a gray, powdered substance that at the slightest touch disintegrated into dust. By using a synthetic resin it was

Figure 9.6 A dark organic mass consisting of remains of faba seeds was found on the bottom of a clay vessel. When it was impregnated with resin and lifted from the bottom, the imprint of the bottom of the clay vessel was clearly recognizable. Radial ribs and a knob made by a knot at the bottom could also be identified, showing that the faba seeds were stored in a sack that was placed within the jar.

possible to consolidate this powder and to identify it as tuffs of barley (*Hordeum vulgare* L.) (Fig. 9.7) similar to those depicted on frescoes and pottery (Fig. 9.8). Some grains of barley were found as impurities in storage vessels in which *Lathyrus* seeds were stored. It is noteworthy that until now barley has been found only in such small amounts. This could have the following explanation: either the harvest was poor or the fleeing inhabitants had been able to take from the settlement large supplies after the first warnings. A third possibility would be that the provisions had already been consumed at the time of the Minoan eruption. Barley, as well as 'faba', were often used as motifs on the pottery found in the excavation at Akrotiri.

In addition to the plants mentioned earlier, one can also identify crocus (*Crocus sativus* Linné) on the frescoes. The pistil of the blossom was collected to make saffron. This was evidently one of the forms of ritual ceremonies that were pursued on the island. This interpretation is based on a fresco showing magnificently dressed and decorated crocus gatherers. This beautiful fresco is shown in Fig. 11.0. Crocus grows on Thera today, and Pliny mentioned Thera as one of the places where saffron was obtained in his time.

The Madonna lily (*Lilium candidum*) is also represented on the frescoes of Akrotiri (Marinatos, 1974; Doumas, 1992). In the so-called spring fresco one sees clusters of red-flowered Madonna lilies on the red volcanic rocks (Fig. 9.9). The characteristic features of the white Madonna lilies are clearly recognizable, but at that time there may have been a red variety of this plant or the artist may have used the color red

Figure 9.7 When a dark powder from the bottom of a small vessel from the Akrotiri excavation was hardened with resin one could recognize the remains of barley.

in order to gain more contrast. White Madonna lilies, however, are also depicted on vases from Akrotiri. From written accounts by Theophrastus (372–287 BC) and Pliny the Elder we know that lilies were used to produce perfume and for use in a medicinal ointment. The ancients raised the bulbs of this species for food and the Greeks and Romans grew it for ornamental and medicinal purposes.

The Madonna lily is a favorite motive in Aegean art, evidenced also by the lilies portrayed on Crete at Amnissos and Knossos, and Hagia Triada and Trianda on Rhodes on wall paintings and pottery. The Madonna lily grows today in northern Greece.

Another plant that can be seen in several frescoes, probably because its cultivation was important, is the Pankratium lily or strand narcissus (*Pankratium maritimum*), which belongs to the Amaryllis family (Diapoulis, 1980; Baumann, 1982). The Bronze Age people of Thera decorated the walls of their houses with it and also used it as a symbol on their ships (Fig. 9.10). However, the representations of it one sees in their frescoes are too large and one wonders whether, instead, it may not be papyrus. The strand narcissus is found today in great numbers near the shores of the Akrotiri peninsula (Fig. 9.11).

The fauna

Animal life is equally well illustrated in the frescoes and pottery of Akrotiri, where artists depicted birds, dolphins, cattle, deer, monkeys, goats, gazelles, and

Figure 9.8 Two white clay vessels from the Akrotiri excavations show motifs of barley (top) and 'faba' (bottom).

Figure 9.9 A, The fresco with lilies and swallows from Akrotiri. B, Part of the beautiful 'spring fresco' shows lilies and flying swallows. Note that the swallows are shown in a perspective view. From the National Museum, Athens, courtesy of Professor Doumas.

Figure 9.10 The ships of the people of Thera were decorated with symbols and flowers as seen in a section of the 'ship fresco' from Akrotiri. (See also Fig. 10.28.) From Marinatos (1974).

Figure 9.11 The strand narcissus (*Pancratium maritimum*) that grows today on Akrotiri peninsula also grew on the Bronze Age island. It is depicted on frescoes from Akrotiri along with tamarisk trees, such as those in the photograph below.

Figure 9.12 One of the best known discoveries of the Akrotiri excavation is the so-called 'fresco of the antelopes'. The two Sömmering gazelles, a type of small antelope, can be identified from the black markings on their faces and from the form of their antlers. Sömmering gazelles live today on the Nile delta and in the arid steppes of Sudan on both sides of the Nile. This fresco, like others showing monkeys, papyrus, ostriches, rivers with palm trees, and the head of an African, is important for archaeologists, because it shows that the people of Thera had contact with the Nile delta of Egypt. From the National Museum, Athens, courtesy of Professor Doumas.

fish. Work on these discoveries is still in progress, but Vanschoonwinkel (1990) has prepared a catalogue of the various animals portrayed in the excavations. He has provided not only identifications of animals found in the excavations but also description of physical remains, such as bones of various animals, including domesticated animals like cows and goats. Even half the skeleton of a pig was among the findings (Marinatos, 1976). There were probably half-wild goats on the island at that time just as there are today on the island of Polyaigos neighboring Santorini (Schultze-Westrum, 1963). The bears among the animals depicted on the island were probably not wild, because the island was too small. But the people of Thera must have encountered wild bears on the Greek mainland or on Anatolia, where they are still seen today.

The exotic elements of the fauna, such as the predatory cat pursuing a stag, the gray or blue monkey, the Sömmering gazelles, and the ostrich egg, indicate that the people of Thera had trade relations with North Africa. This interpretation finds further support in the portrayal of Africans on the frescoes. The monkeys seen in the frescoes may be related to those still living today on the rocks of Gibraltar. The Sömmering gazelle provide a link to the Nile delta (Fig. 9.12). In the first years of the excavations of Akrotiri, Spyridon Marinatos referred to North Africa (Libya) as the region of origin for these exotic discoveries. Since then, his interpretation gains added support from the recent discovery of Minoan frescoes on the Nile delta, one of which dates from the time of the period of Hyksos and the beginning of the New Kingdom (Bietak, 1992; Bietak and Marinatos, 1993).

10

An idea takes hold

Recent geological and archaeological discoveries enable us to reconstruct the island as it was before the eruption. Rocks and minerals from the excavations offer insights into the geological relations of the original volcanic island, and we can learn much about techniques used in constructing buildings and making pottery. We are also able – in some cases – to reconstruct the trade routes used at that time.

Figure 10.0 Important information about the geology of Santorini is stored in the light-colored limestone blocks (stromatolites) that lie on the slopes of Megalo Vouno near the northern rim of the caldera. They were thrown out in the third phase of the Minoan eruption. Formed in shallow sea water of the flooded caldera around 12 000 years ago, the limestone contains fossils of marine animals that prove that a caldera existed prior to the Minoan eruption in 1640 BC.

Geological clues for reconstruction of the island

What was Santorini like before the Minoan eruption? Was the island round, or did it have quite a different form? Did the volcano have a high summit at its center, as was thought for some decades (Luce, 1973), or did a water-filled caldera already occupy the center of the island? Questions such as these were the central theme of the Third International Thera Conference held on Santorini in September 1989. Scientists representing various fields of study and viewpoints contributed to the common theme of Thera and the Aegean World. The geological discussions were devoted to the form of the pre-Minoan island.

During the second Thera Congress, which was held in August 1978, we could already show that before the Minoan eruption the island had not been perfectly round (Pichler and Friedrich, 1980). It had a water-filled inlet that cut into the southwestern side (Fig. 10.1A). At that time we accepted the view that before the Minoan eruption the island consisted of numerous hilltops of rather modest size and elevation. We suggested that the island could have had an inlet that served as a harbor during the Bronze Age. This reconstruction was based mainly on interpretations of the Minoan eruption and the morphology of the present islands (Fig. 10.1).

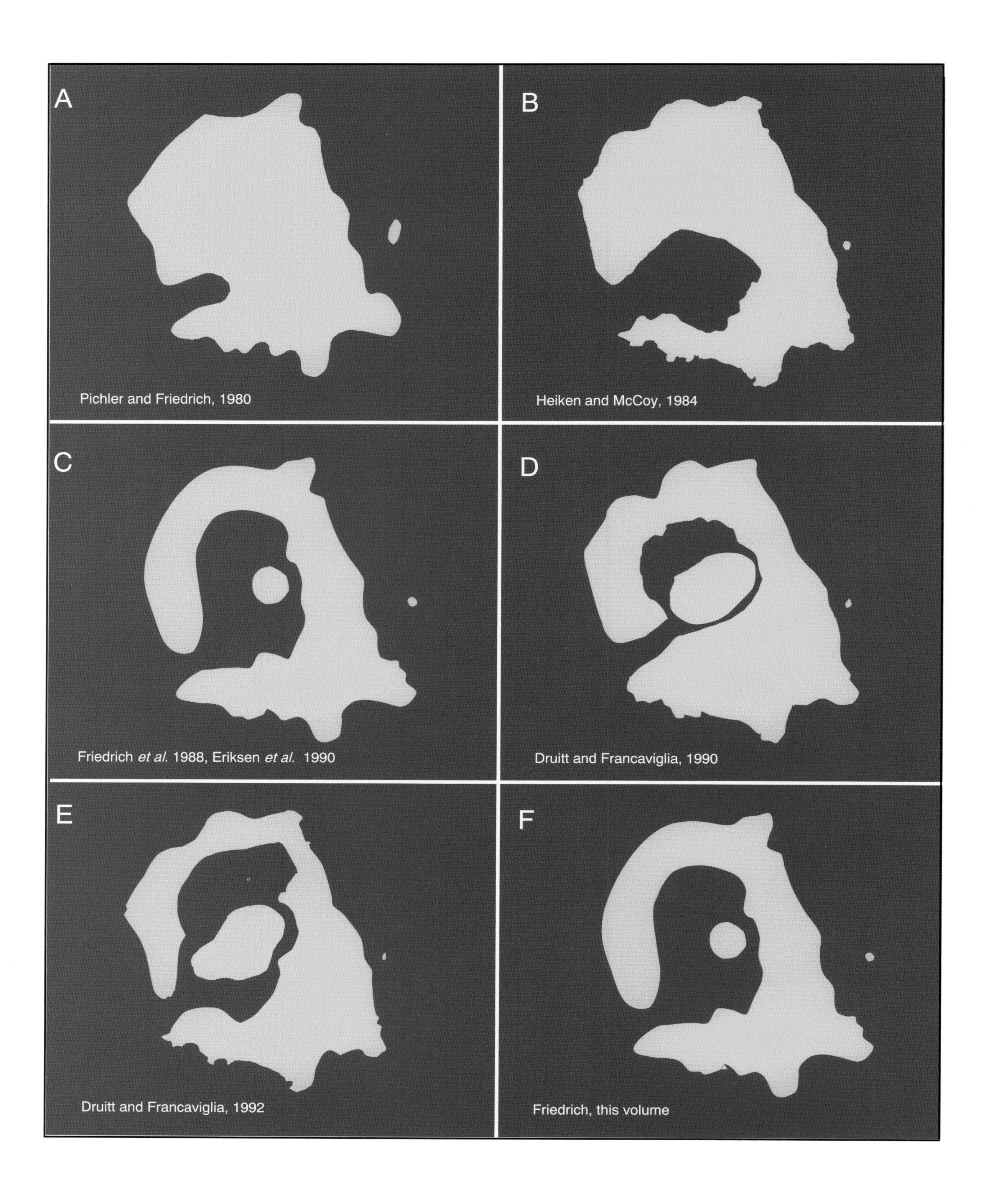
A
Pichler and Friedrich, 1980
B
Heiken and McCoy, 1984
C
Friedrich *et al.* 1988, Eriksen *et al.* 1990
D
Druitt and Francaviglia, 1990
E
Druitt and Francaviglia, 1992
F
Friedrich, this volume

Figure 10.2 The light-colored blocks of limestone (stromatolites) are conspicuous when surrounded by darker stones in a wall.

Six years later the American geologists Heiken and McCoy (1984) offered a graphic portrayal of the island before the Minoan eruption. They had carried out a comprehensive survey of the Minoan pumice layer on Thera, Therasia, and Aspronisi and subtracted the measured thickness of this layer from the surface elevations shown on a topographic map. In this way, they could reconstruct the land surface of the island before the eruption. The caldera deduced from their reconstruction was larger than in our earlier reconstruction of 1978 (Fig. 10.1B).

Figure 10.1 (Opposite) The progress of geological research on Santorini can be illustrated by the six reconstructions published since 1980. In reconstruction A, the embayment in the southwest is shown; in the east the small Monolitos island is visible. This small island and the opening of the caldera are shown in all following reconstructions. B shows a larger basin in the southwest. In C the southwestern and northern basins of the caldera are flooded and the Pre-Kameni island is located near the center. In D the caldera walls at Fira and Megalo Vouno are drawn in more detail and the central Pre-Kameni island is larger than in C, while the caldera wall in the area of Athinios is still visible. E shows that an opening to the southwest existed before the Minoan eruption and that the Akrotiri Peninsula had a similar shape as today. F as in C.

This was not the end to the story, however. Another reconstruction of the island came ten years later, but from a quite unexpected source. Exotic blocks of an unusual type of limestone were found in the pumice of the Minoan eruption. Being as light-colored as the pumice in which they are embedded, they are not normally conspicuous, but they immediately attract notice if they are in a wall built mainly of black lava (Fig. 10.2). In 1871, Karl von Fritsch had noted these white blocks of limestone and, judging from the fossils identified in them by K. Mayer, reasoned that there must have been a shallow body of water near the vent, possibly a lagoon (see Box 10.1). The fossils were recognized as very recent forms. Fouqué (1879) also commented on these blocks and agreed with von Fritsch's interpretation. It was not until 120 years later, however, that their importance would be recognized. Their presence in the pumice meant that at the time of the eruption they were torn from the walls of the volcanic vent and thrown out together with an enormous mass of pumice. Their approximate age was already known; Quenstedt (1936) described fossils from these blocks and placed them in the 'Late Quaternary'.

Box 10.1 Karl von Fritsch's observations on the limestone blocks of Santorini

'The blocks of this limestone, which may have a dark color owing to the bits of lava they contain, are found in the upper white tuff on Therasia and on Thera near Apanomeria [Oia], Phinikia, Megalo Vouno, Vuvulos, Fira and Mesaria and perhaps even farther to the south. There is no doubt that these blocks are within the tuff. Their wide distribution means that strong explosions disrupted their original deposits. Fossils are occasionally found in this limestone, and Dr. K. Mayer identified the most common forms as Bythinia ulvae *Penn sp. (Turbo). A small piece of this limestone from Therasia also contained what Dr. Mayer identified as* Cerithium conicum *Blainv. (mamillatum Risso). Both of these forms indicate a rather recent age, since both are still living and, in fact, Cerithium is considered diagnostic of limestone formed in salty or brackish water, possibly a lagoon. Whether the layer of limestone was associated with the late Tertiary tuffs of the Akrotiri area is yet to be determined.'*

In the course of the studies of Santorini that I and my colleagues have carried on since 1975, I have noticed many of these blocks and recognized them as a type of algal limestone known as stromatolites (see Fig. 10.3 and Box 10.2). Eventually, in 1987, one of my students at Aarhus, Ulrike Eriksen, undertook a study of them for her graduate thesis. She quickly developed a special ability to spot them among the blocks used by farmers for the walls of their vineyard terraces. She found many more in quarries on Thera and Therasia, where they had been sorted out as waste and left in piles. By concentrating on the stone walls and quarries, it was possible to find large numbers of these blocks and measure their dimensions. Altogether, Eriksen recorded 1406 blocks, with a total weight of 56 tons. This, of course, was only a very small fraction of the total amount of this material that must have been ejected, but it was enough to resolve the question of where they came from. By mapping their distribution by size, she hoped to find the location of the vent from which they were ejected. According to the laws of ballistics, the largest blocks should be found closest to the vent. As we expected, they showed that their source was in the northern part of the present caldera.

Figure 10.3 Blocks of limestone are found in the pumice of the third phase of the Minoan eruption. This block on the caldera wall at Micro Profitis Elias clearly shows the concentric layering of the stromatolites. The coin is about 2 centimeters in diameter.

This confirmed the conclusion of Pichler and Kussmaul (1972), who had earlier proposed that the focus of the eruption 'must have been in the vicinity of the present Kameni Islands'. The distribution of limestone blocks allowed us to locate that point more precisely: the vent must have been northeast of Nea Kameni (Friedrich *et al.*, 1988; Eriksen *et al.*, 1990). In this area we now find an arcuate escarpment of the caldera which reaches a depth of nearly 400 meters below sea level (Fig. 10.4).

Stromatolites are normally formed in shallow water where the mats of algal bacteria can obtain light, but we had to know whether the limestone was formed in fresh or marine water. If they were formed in salt water, the caldera must have been flooded by the sea. The question was resolved when Bjoern Buchardt analyzed the oxygen isotopes of the limestone and Ulrike Eriksen found thousands of small snails and other fossils in some of the blocks: they were formed in marine conditions, as will be shown later (Fig. 10.5).

Fossils solve the problem

We were also anxious to know the age of the limestone blocks. Quenstedt (1936) had already determined that they were from the Late Quaternary, but it might be possible to date them more precisely by the radiocarbon method, provided they were younger than the upper limit of the method, 50 000 years. In the autumn of 1987 I sent a block selected by Eriksen to Henrik Tauber at the radiocarbon laboratory in Copenhagen, and early in 1988 the first results were received. Our expectations were confirmed: the limestone was datable and had a radiocarbon age of about 18 000 years. We gave part of the same block to Mette Skovhus Thomsen in Aarhus for a radiocarbon investigation using the AMS-method. As we hoped, it gave the same age; however, the values of the delta ^{13}C measured in Copenhagen were extremely high. These measurements of the stable ^{13}C isotope gave a value of 12, while 1 is normal. This could be explained by the fact that the stromatolites had grown exactly above the

Box 10.2 Stromatolites

A stromatolite is limestone formed by mats of algae or bacteria. Normally, gastropods feed on these mats, but if conditions, such as the salinity and temperature of the water, are favorable, the mats of algae or bacteria grow and form stromatolites. They can trap sediments on their surface and can precipitate calcium carbonate in the lower parts of the mats.

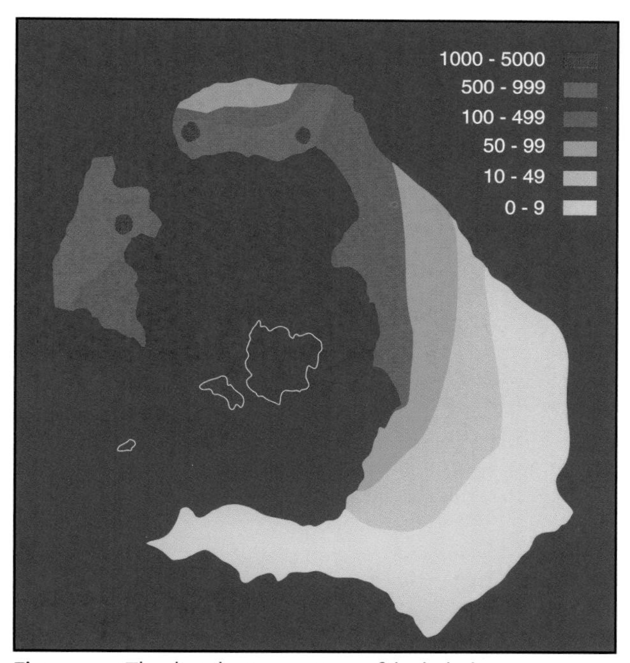

Figure 10.4 The distribution pattern of the light limestone (stromatolites) on Thera and Therasia shows a clear concentration in the northern parts of the two islands. Measurements of 1406 limestone blocks have been entered into a computer program that sorted them and plotted contours of their weight in kilograms. The diagram shows that heaviest and largest blocks are found in the north, while they decrease in size away from the caldera. This pattern shows that the area of formation of the stromatolites and also the vent of the Minoan eruption were located within the region of the present caldera in an area north of where Nea Kameni is today. This part of the caldera is nearly 400 meters deep. (Modified after Eriksen, 1990.)

Figure 10.5 The thousands of fossils found in cavities in the stromatolites are only a few millimeters in size. They are mainly gastropods that lived in the shallow water of the caldera before the Minoan eruption. They lived on bacterial algae mats. Photo by U. Erikson (1990).

magma chamber, where they could absorb older carbon dioxide gases from the volcano dissolved in the water. This meant that the measured radiocarbon age of about 18 000 years was too old and had to be reduced. This was confirmed when isolated snails were measured using the AMS-method. They gave younger ages (Table 10.1). Bjoern Buchardt in Copenhagen later used the same material for a study of the stable oxygen and carbon isotopes. His results confirmed our earlier idea that the limestone blocks were formed under marine conditions.

In the meantime, Eriksen had found other clues indicating that the limestone was formed in marine water: she was able to identify the small snails found in the limestone blocks as brackish marine forms. The same species still live in the Mediterranean today: there were two fossil species, *Hydrobia neglecta* Muus and *Pirenella conica* (Blainville), as well as the ostracod *Cyprideis turosa* (Jones), foraminifera *Ammonia beccarii* var. *tepida* (Cushman), and numerous algae with silicified skeletons (diatoms). These fossils gave us further information about the source area of the limestones: they were from shallow sea water. It is known that the two snails and ostracods can exist under conditions of extreme salinity and temperature, but they need normal marine conditions to propagate.

More pieces added to the reconstruction

Now we knew for sure that the algal limestone blocks were formed in the seawater-flooded caldera about twelve thousand years ago and that they were thrown out in the final phase of the Minoan eruption. The caldera already existed in the Stone Age. We pub-

Table 10.1 *Radiocarbon ages from Santorini based on samples of stromatolites, gastropods, marine travertine, and charcoal of trees that were covered by an ignimbrite*

Laboratory	Sample	Years BC	^{14}C (ppm)	Material
Radiocarbon laboratory Copenhagen	K-5104	16 750 ± 270	12.3	Stromatolite
Radiocarbon laboratory Copenhagen	K-5367	19 470 ± 280	13.8	Stromatolite
AMS University of Aarhus	*AAR-14	17 730 ± 240		Stromatolite
AMS University of Aarhus	*AAR-15	18 360 ± 250		Stromatolite
AMS University of Aarhus	*AAR57	21 600 ± 290		Marine travertine
AMS University of Aarhus	*AAR-58	20 800 ± 330		Marine travertine
AMS University of Aarhus	*AAR-59	17 900 ± 250		Pirenella gastropods
AMS University of Aarhus	*AAR-60	17 600 ± 220		Pirenella gastropods
AMS University of Aarhus	*AAR-61	13 010 ± 160		Hydrobia gastropods
AMS University of Aarhus	*AAR-62	12 860 ± 170		Hydrobia gastropods
AMS University of Aarhus	*AAR-37	18 150 ± 200	−25	Charcoal in ignimbrite

The data are partly from Friedrich *et al.* (1988); those marked by an asterisk are from Eriksen *et al.* (1990).

lished our observations about the flooded caldera in 1988, and shortly after, at the Third Thera Congress, our British and Italian colleagues, Tim Druitt and Enzo Francaviglia (1990), reported exciting observations that confirmed our results. They had found *in situ* pumice of the Minoan eruption 'plastered' on the inner side of the caldera at a height of about 130 meters. They also observed that the caldera wall at Katofira was strongly weathered. They concluded from this that a large part of the caldera had existed before the Minoan eruption (Druitt and Francaviglia, 1990). This added an important piece of evidence for the existence of the caldera before the Minoan eruption. Heiken and McCoy (1984) had made a similar observation in the Megalo Vouno area where they observed that the uppermost third of the caldera wall was deeply weathered. Further evidence for our new caldera theory was found in other places on the caldera walls, where the Upper Pumice (Bo) dips into the caldera (Fig. 10.6) (Friedrich *et al.*, 1990).

Minerals for color pigments

During the Thera Congress in September 1989, one of the archaeologists reported that the white clay and colored pigments used in jars at Akrotiri had been identified by means of an electron microscope (Vaughan, 1990). The white pigment was found to be the mineral talc (Fig. 10.7).

Since outcrops of talc on Santorini were unknown to the scientists at that time, it was concluded that it must have been imported. When I heard this, I immediately thought about the talc I had seen at an old mine pit on the inner side of the caldera at Cape Plaka (Fig. 10.8). The owner of this area, Parthenios Gavalas, had showed it to me. In addition to talc there were other minerals that could have been used as color pigments. White, red, blue, and green minerals are found in the phyllites of the metamorphic series close to the warm springs in the caldera wall between Cape Plaka and the harbor at Athinios (Fig. 10.9) and close to the church at Cape Therma. The blue and green minerals are the copper-bearing minerals malachite and chrysocolla. The sites are all on the inner side of the caldera within 50 meters of the present sea level. We obtained analyses of minerals found at Cape Thermia during a field trip with students (Klitgaard, 1986; Fig. 10.10).

At present my colleagues at Aarhus Sidsel Grundvig and Svend Erik Rasmussen are investigating whether minerals from Cape Plaka and Cape Therma were

Figure 10.6 An important clue for the reconstruction of the Bronze Age caldera is still visible on the present caldera walls. On northern Therasia, the layers of the Minoan eruption dip deeply into the caldera, as is seen in the foreground on the left.

used as color pigments for the coloring of frescoes at Akrotiri.

This discovery led to further questions and speculations. If these minerals were already accessible to the Bronze Age people, then other evidence should also be re-examined. Once it was clear that the most of the flooded caldera existed before the Minoan eruption many new questions arose. Were there other important mineral pigments close to the present caldera wall? Were the lead and silver ores in the caldera walls at Athinios accessible at that time? We formulated a short 'addendum' to address these questions (Friedrich and Doumas, 1990).

The rocks, minerals, and soil found in the excavations at Akrotiri helped us to add more pieces of the picture of the Minoan geology and landscape. If one

Figure 10.7 Two white vases from the excavation of Akrotiri are decorated with breasts and bird's necks. The white pigment on the one on the left is a thin coat of talc. This mineral is found today on the inner side of the caldera at Cape Plaka where the pits from which it was mined are still visible. It was probably accessible during the Bronze Age, because the Minoan deposits on the caldera wall lie on the steep slope that dips into the caldera.

assumes that in building the settlement at Akrotiri most of the material was gathered from the immediate neighborhood, then the stones used in houses give us additional clues to the geology of the island and sources of building materials and even to possible trade routes. We can assume that heavy loads were transported by boat and only lighter items were carried by men or animals. But what was the former coastal region like at that time? In a few places where the Minoan pumice is already stripped away it is possible to recognize the Bronze Age shore and former quarries. According to Flemming (1986), the sea had about the same height as today. But we should consider that the coastal areas could have been cut back as they were during the eruption of 1707 to 1711 (see Chapter 12). If so, the coast line in the Bronze Age reached a few meters farther into the sea. This can be confirmed by the submarine platform consisting of beach rocks that is parallel to the present shoreline south of the Akrotiri peninsula. A similar platform formed of phyllite and marble is found on the inner side of the caldera at Cape Plaka, where it forms a seven-meter deep calcareous layer about 40 to 50 meters wide. If this shallow shelf was above the sea in the Bronze Age, the shoreline was somewhat farther into the caldera.

This locality at Plaka is also the place where talc and colored minerals were probably accessible before the

Figure 10.8 The white mineral talc is found in the phyllites at Plaka on Thera. An old talc mine is still visible in the caldera wall at a height of about 30 meters above sea level. The white area in the foreground is the waste-pile, but on the left the outcropping rock is *in situ*. The houses of Plaka are seen in the background.

Minoan eruption. Two geological observations support this inference. The phyllites of the metamorphic basement dip steeply in toward the caldera. At the coast close to Plaka they dip down at an angle of 43 degrees toward the west. The discordant overlaying volcanic series dips in the same direction, and the same is true of the mineral deposits. Beneath the two windmills standing on the caldera rim at Megalochori the Upper Pumice (Bo) layer dips 4 to 7 degrees west (Fig. 10.11), and on the caldera walls between Plaka and Athinios the angle is even steeper – 22 degrees. Druitt and Francaviglia (1992) measured 11 degrees at Athinios. The deposits of the Minoan eruption were therefore emplaced on this part of the inner caldera wall where they follow the topography of the underlying surface; this shows that there was a basin in this area (Figs. 10.11 and 10.12).

One rock that has certainly been imported is obsidian. When freshly broken, this dark volcanic glass has sharp edges that make it useful for tools and weapons. Obsidian used for these purposes is found on the old surface on Thera and Therasia. Geologists and archaeologists agree that it came from the island of Milos where prehistoric obsidian mines have been found (Renfrew, 1985).

Figure 10.9 This cave is located about 200 meters south of the harbor entrance at Athinios and 50 meters above sea level. About 7 meters deep, it contains several blue, green and black minerals. The black is due to coatings of manganese oxides on fracture surfaces, and the blue and green are from copper minerals. In places the latter are up to 2 centimeters thick. The dark red color of the rocks is from iron oxides that were probably carried by the wind from the iron-rich precipitates on the shores of the Kameni islands.

Figure 10.10 Only a few meters above sea level on the beach at Cape Thermia there is a small exposure of strongly weathered phyllites containing blue minerals. The mineralization at this locality was mentioned by Fouqué as early as 1879. According to Wilski (1934) the deposit was once mined for lead and zinc.

Stones and building materials of the Theran people

The inhabitants of the Bronze Age settlement at Akrotiri built their houses using materials found in their immediate neighborhood (Fig. 10.13). This is the impression one gets when visiting the excavation. The rocks are quite gray, because a thick layer of dust covers everything, but outside the excavation where some of the rocks are stored in the open their natural colors are visible. Two colors are predominant: the red of volcanic rocks (ignimbrites) and the gray of the fine-grained tuff used to make rectangular blocks. These rocks were used in houses that seem to have been important because they contain unusual cultural items and offering bowls. Elsewhere, the gray rock was used only for the corners, doorways, and window frames. Most houses were constructed of cruder masonry with inserted branches of wood that gave the buildings greater stability in the event of earthquakes.

The 21 000-year-old red ignimbrite mentioned in Chapter 3 is exposed at Potamos, a short distance east of the excavations, and reaches the sea at a site halfway between the Akrotiri excavations and Vlihada. Furthermore, it is also seen at the base of the Akrotiri excavations at pillar 17, where it formed the roof of a

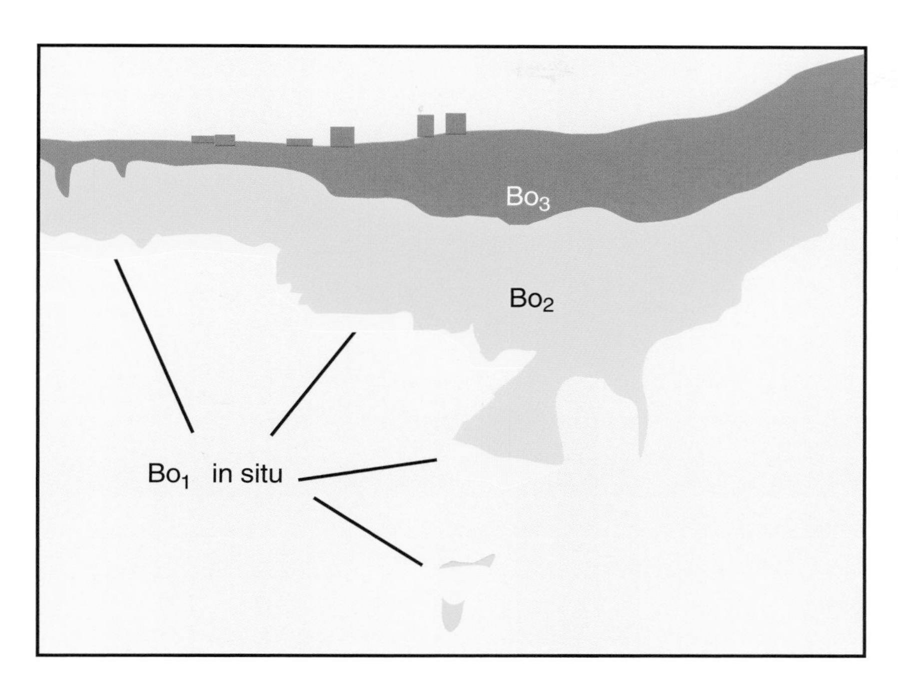

Figure 10.11 On the caldera wall beneath the town of Megalochori (photo, bottom) the light pumice layer of the Minoan eruption shows a distinct inclination into the caldera. The angle of dip is about seven degrees to the west. The diagram above shows the original position of the Bo_1 layer.

Figure 10.12 The discordance between the metamorphic and volcanic rocks is clearly visible at Athinios on Thera. The volcanic deposits mantling the older topography were deposited on a sloping surface of basement rocks. Here we see a depression in the pre-volcanic topography. Today the tuff left by erosion resembles patches of wallpaper on a wall (in the very center of the photograph). The Minoan pumice here is inclined downward 11 degrees to the west.

cave. Both at Akrotiri and at Potamos some of the houses were erected directly on this ignimbrite. Before the Minoan eruption it formed a platform inclined 8 degrees toward the south and filled the former erosion channels at Potamos and Akrotiri.

The light gray tuff (A1-tuff of Pichler and Kussmaul, 1980a), which was used not only in construction but also to make Minoan bull horns, is found near the modern village of Akrotiri. Furthermore, it occurs along the coast at a place about one kilometer west of the Bronze Age settlement that is often referred to as White Beach. The latter locality is probably where the building stone was obtained,

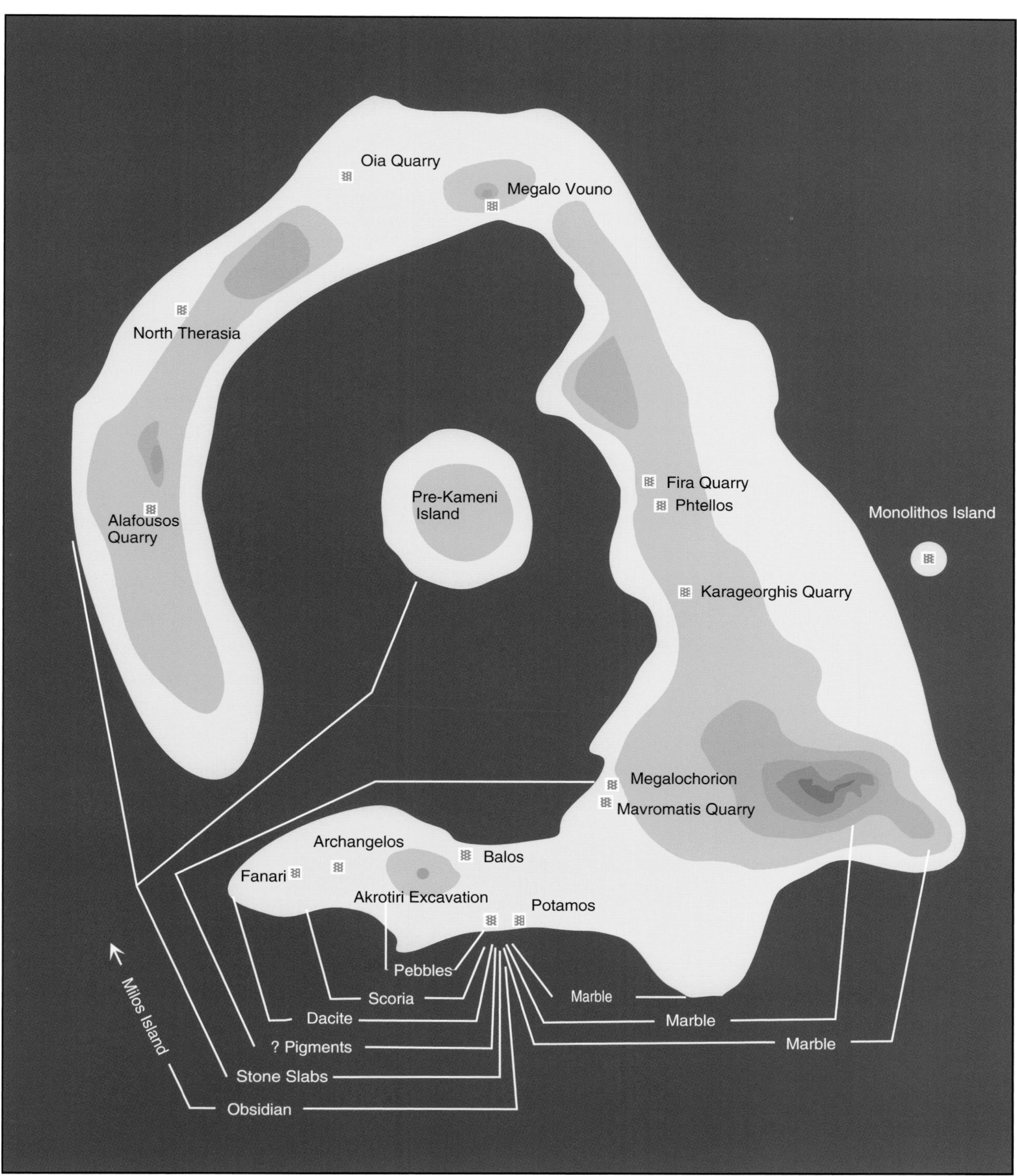

Figure 10.13 Possible sources of rocks, minerals, and clay are shown in this reconstructed form of the Bronze Age island. Some of the information is from Einfalt (1978b). Squares mark Bronze Age sites. The height contours have a spacing of 100 meters.

because it is found lying on the beach today as it probably was in the past.

The floors of the rooms in the Akrotiri excavations were formed of plates of lava about 4 centimeters thick. These plates were used especially in the upper floors. Holes in the walls can be seen where the beams supported these floors. These beams were covered with earth and then with mats of Italian reeds and finally with a surface made of irregular slabs of lava. A quite similar style of floor construction was used in the Bronze Age houses at Chania on Crete (Hallager, 1990).

Until now, no one has determined the source of the lava slabs for sure. Such platy lava is found in the lava domes of Palea Kameni and in the Skaros sill (Einfalt, 1978b). On Micro Profitis Elias it occurs in platy dike rocks. On Therasia it is found on the southwest coast at Cape Camina, where an updomed, onion-shaped structure is exposed in cross section. The platy structure seems to have been produced by shearing the magma before it was completely solid (Chapter 1, Fig. 1.5).

While Palea Kameni has to be excluded because it was formed after the Minoan eruption, the areas of Skaros on the inner side of the caldera and the sites on Therasia could be possible sources for this material, because they were probably accessible from the sea in the Bronze Age. Another possibility could be the Pre-Kameni island, because it was presumably similar to Palea Kameni and was situated in the part of the caldera that was accessible at that time.

Not only did the people of Thera produce sophisticated ceramics; they were also able to make hammers, anchors, millstones and even stone vases. Most of them are made of local volcanic stone, like the more than one-meter high Middle Cycladic stone jar from Akrotiri (Doumas, 1983, fig. 17). But in some cases it is still uncertain where the material came from. In the case of an unfinished stone vase, made of reddish marble, it was possible to study the production process in detail (Warren, 1978), but the source of the material could not be identified with certainty. Warren (1978) speculated that the red marble could be rosso antico from the quarries at Mani in southern Laconia, but comparative analyses of the stone by Einfalt were inconclusive. The incomplete vase measures 57 by 55 centimeters and the block from which it was being made weighed about 200 kilos. The fact that the bottom of the vase was not yet finished indicates, in my opinion, that it is a local product (Fig. 10.14). How could that be proved? A similar rock is found on Thera at Gavrilos close to the tombs at Echendra. In September 1993, I received a tiny sample of the stone and had it analyzed by Bjoern Buchardt in Copenhagen. The analysis was found to fall within the range of samples from Echendna, showing that the material of the vase could be from this source area. More samples should be analyzed, however, since the rocks at Echendra vary widely. The stone vase was made of impure marble, consisting of about 80 percent calcium carbonate by weight, which foams when tested with a drop of hydrochloric acid. When its colors were determined by means of a color chart, the four components (cyan 0, magenta 40, yellow 50 and black 10) were found to be the same as those of the rocks at Echendra (Fig. 10.15).

Einfalt also noted that pebbles of reddish marble found in the excavation at Akrotiri could be derived from the Gavrilos area, where steep south-dipping marble and phyllites are exposed. The locality is bounded by a platform of pumice, which now reaches the sea but which did not yet exist before the Minoan eruption. Thus, the southern part of Gavrilos was formerly situated close to the shore and was accessible from Akrotiri by boat (Fig. 6.10).

The gray-green marble found at Akrotiri could have been obtained from exposures on the slopes of Profitis Elias, such as the region between Emborion and Perissa noted earlier by Einfalt (1978b). Material appears to have been excavated from this site at some time in the past. Moreover, this place was situated directly on the shore before the Minoan eruption and could have been reached by boats from Akrotiri (Fig. 10.16).

Figure 10.14 An unfinished stone vase found at Akrotiri was made by a special drilling technique used to carve the reddish rock. The rock appears white in the photo because it has a light covering of dust. The source of the rock is uncertain, but chemical analyses show that it could have come from the area of Echendra on Gavrilos Ridge.

Good plastic clay is unknown today on Santorini, but Einfalt's studies (1978a) indicate that some pottery sherds from the excavation at Akrotiri could have been made from clays that came from Thera.

Fouqué (1879) and Phillipson (1896) had already pointed to the weathered phyllites as a possible source of pottery clay deposits. These are found in several places, including the northern side of Cape Athinios (Neumann van Padang, 1936), the steep slopes at Plaka and Thermia, and the saddle between Profitis Elias and Mesa Vouno. Fouqué found small fossils in some of the pottery sherds he collected from the area around Akrotiri and concluded that the Late Pliocene marine sediments on the Akrotiri peninsula had been used for some of the ceramics. This deduction was confirmed a few years ago when clay was found in a hole dug for a pillar to support the roof over the Akrotiri excavations (Marinatos, 1976). Vaughan (1990) showed that, when heated to 900 °C and examined under the microscope, the clay was very similar to that used in Minoan ceramics of Akrotiri. Thus, she showed that pottery that archaeologists said was of local origin had been made from this clay. In July 1992 I took a sample of this clay from pillar 17 and asked Marit-Solveig Seidenkrantz to examine the fossils it contained. She found in this sample an assemblage of foraminifera similar to that from a locality at Archangelos on Akrotiri peninsula, removing any doubt that the clay in the excavation is from Thera (Seidenkrantz and Friedrich, 1992). Until the excavations are completed it will be impossible to say whether the clay at this place was formed *in situ*, or whether it had been washed in over the course of time by the strong winter rains. It is also possible that the people of Thera had deposited the clay at this place, but this seems less likely, because the layering in the clay slopes down from the entrance to a cave beneath the Cape Riva ignimbrite. Sediments may have been washed in from the Lumaravi–Archangelos area by rainwater coming down the sloping surface of the ignimbrite and may have collected in natural or man-made depressions. This possibility was confirmed by analyses performed at Aarhus by Ole Bjørslev Nielsen (Friedrich *et al.*, in press).

Rocks shown on frescoes

On the lowermost part of some frescoes at the Akrotiri excavation one can see rectangular layers painted to resemble natural rocks. The artist was obviously trying to depict some of the rocks on Santorini. This is especially clear in a fresco from the so-called 'West House'. The colors and clear pattern of folds closely

Figure 10.15 At the Gavrilos ridge at Echendra several tombs were carved into the reddish marble-like rock during Hellenistic times. The unfinished vase at Akrotiri was probably made from rock from this locality. The area was situated close to the shore before the Minoan eruption and was accessible from Akrotiri by boat.

resemble those of phyllites that have been subjected to compressional folding. Rocks of this kind can be seen in the neighborhood of the hot springs at Cape Plaka (Figs. 10.17 and 10.18). The same fresco shows also a painted rock resembling a weathered volcanic rock with concentric structures (Fig.10.19).

Two of the frescoes found at Akrotiri are especially noteworthy. The 'spring fresco' (Fig. 9.9) showing red lilies and swallows was among the first discovered at Akrotiri. Another shows girls gathering crocus (Fig. 11.0). In both of these frescoes one can see rocks that are generally considered to be lava (Marinatos, 1974).

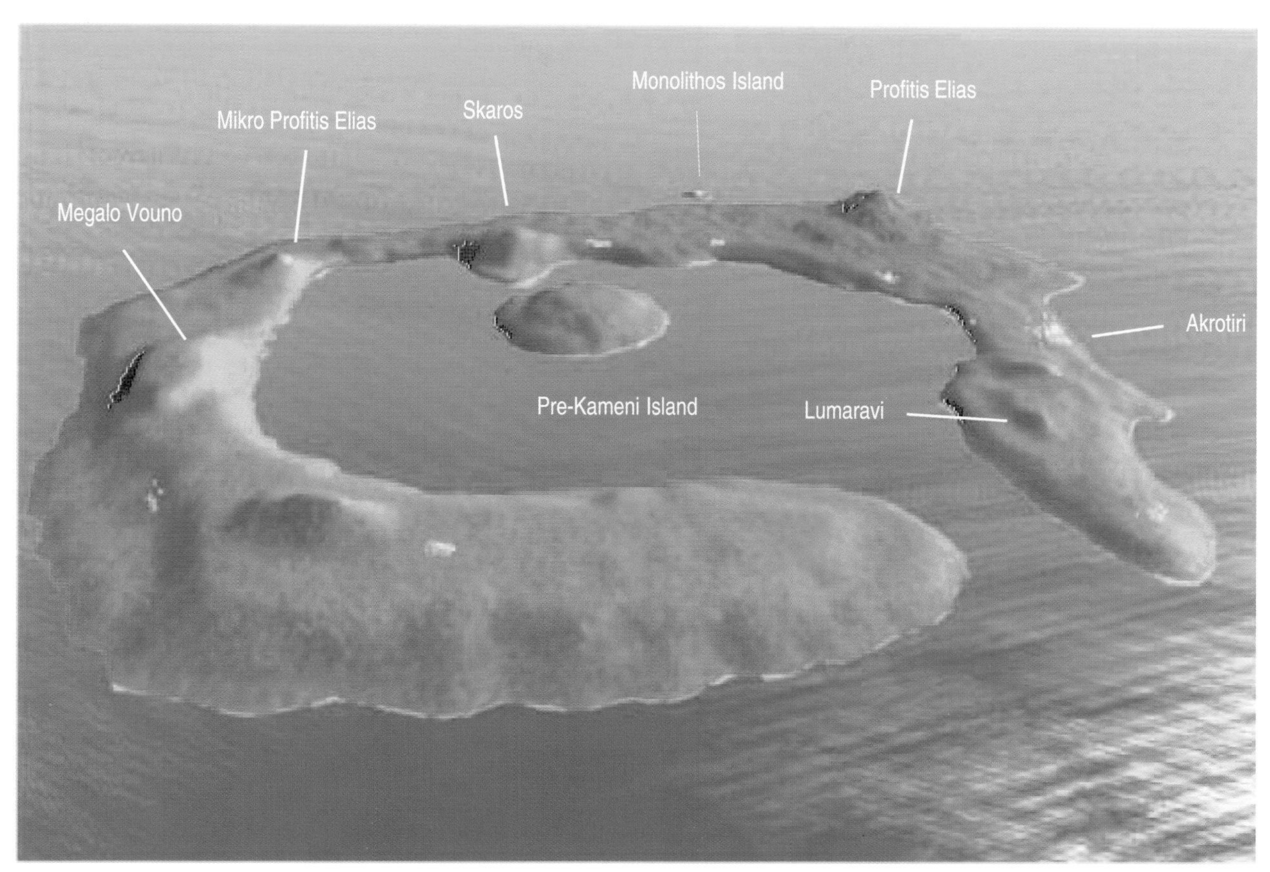

Figure 10.16 A reconstruction shows the Bronze Age ring-island in a bird's-eye view from the west. In the foreground one sees the reconstructed parts of Therasia and Aspronisi that were formerly connected to Thera. The pre-Kameni island is situated in the middle of the caldera, and Akrotiri is near the shore on the far right. White dots mark the known Bronze Age sites. The reconstruction is based on numerous observations and measurements of the Minoan deposits as well as the topography of Thera, Therasia, and Aspronisi.

They resemble rocks in the vicinity of Akrotiri close to the beach at the Nicolas church. Quite similar rocks occur on the inner side of the caldera on Akrotiri peninsula at Kokkinopetra, which means red rock.

Metallurgy

Among the bronze artifacts found at Akrotiri are knives, lances, and scythes, as well as pans and kettles, all of which show a high level of metal work and practical design. The copper and tin used in making these implements were probably not found on Santorini. More likely they were obtained by trade, possibly from Cyprus and Laurion.

Archaeological evidence for the reconstruction of the ring-island

When growing geological evidence showed that a water-filled caldera existed prior to the Minoan eruption, it opened a new perspective of the Bronze Age settlement. Until then, it was thought that the inner part of the island was a high landmass; even as late as in the 1970s a number of books and tourist guides stated that the central volcano had a height of 1600 meters. Bronze Age buildings that, over the years, had been discovered in the pumice quarries near the walls of the present caldera had been assigned by archaeologists to Bronze Age shepherds, but after the new caldera concept became evident these interpretations were revised.

Figure 10.17 The fresco from room five of the West House in the excavation of Akrotiri shows a small cabin standing on a base of stone. Similar cabins are shown with human occupants in the ship fresco (Fig. 10.26). The rocks are painted in such a realistic manner that they can easily be identified. In my opinion, the two main rocks of Santorini are depicted here. On the left is the non-volcanic rock (phyllite) and on the right volcanic slag with concentric layering (see Figs. 10.18 and 10.19). From the National Museum, Athens, courtesy of Professor Doumas.

Old settlements on the caldera walls

In the nineteenth century, when pumice was being mined near the town of Fira, two carved marble figures and a few bowls made of white marble were found in a quarry near the town of Fira. These objects could possibly be from tombs. They are examples of the art of the islanders at the time of early Cycladic II. The two statuettes are now in the Badisches Landesmuseum in Karlsruhe. Unfortunately, the circumstances of their discovery are unknown but

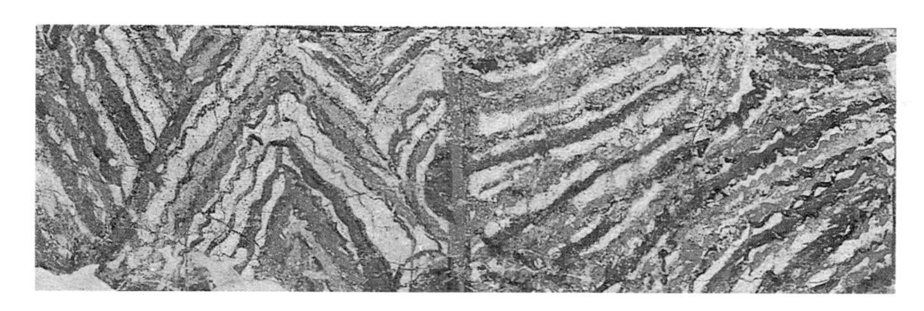

Figure 10.18 This illustration shows a comparison of the painted rocks shown in the fresco above with real rocks of Santorini below. The characteristic compressive folding of the phyllite is seen in both.

Figure 10.19 Volcanic rocks, such as this weathered tuff from Cape Plaka, were probably depicted on the frescoes. The concentric structures of the volcanic bombs resemble those seen in the fresco.

according to museum officials they are from the collection of Friedrich Maler, who obtained them in 1840 (Fig. 10.20).

Tombs were found at the end of the nineteenth century in the quarry at Fira (Karo, 1930). One of these contained a small figurine or 'idol' and a bowl carved from steatite, a very soft rock. Karo supposed that this quarry was the site of an important cemetery. The tombs were directly adjacent to the caldera wall. They were covered with stone plates of 'limestone from

Figure 10.20 The two marble harp players show a high level of artistry around 2500 BC. They were probably found in the quarry south of Fira where tombs were discovered in the last century. Photo courtesy of the Badisches Landesmuseum, Karlsruhe, Germany.

other parts of the island'. Presumably, this was limestone of the type found today on the inner wall of the caldera at Cape Athinios and Plaka. The geologist Hans Reck (1936) also mentioned that tombs were found in this quarry. Other ceramic remains were often found as work continued during the twentieth century. The Greek geophysicist Galanopoulos (1958), for example, reports on archaeological findings in tombs under the pumice of this quarry (Box 10.3).

In 1975 I was carrying on geological investigations in this same quarry. One could see remains of a paved path and painted pottery and the remains of a Cycladic house. I could not determine whether this was the same place which Galanopoulos had seen earlier, but the ruins were fenced in, and they were later removed by subsequent mining.

In May 1980, together with my twelve-year-old son Andreas and my colleagues Rud Friborg and Birthe Schmidt, I was investigating the ash layers in the quarry south of Fira. While I was measuring the profile, Andreas was playing on a debris slope at the base of the profile. Suddenly he called me and showed me a pottery shard (Fig.10.21).

Box 10.3 Archaeological discoveries in the quarry near Fira by A. G. Galanapoulos

'While investigating the consequences of the Amorgos earthquake on the 9th of July, 1956, I received a message from Mr. Papageorgiou telling me that ruins of walls and prehistoric objects had been found in the pumice quarry close to Fira. During my visit to this quarry, I succeeded in finding close under the Rose Pumice layer and immediately on the deeply weathered soil to discover a nearly vertical, one- to two-meter wall, humans bones and teeth, as well as fragments of painted pottery and charred tree bark. Soon after my return the charcoal was sent to the Lamont Geological Observatory, Columbia University, Palisades, New York, for a ^{14}C dating. The charcoal (L 362) first gave an age of 3050 plus or minus 150 years (Galanapoulos, 1957), but a new measurement of the same sample gave 3370 plus or minus 100 years.'

Figure 10.21 During geological studies of the quarry south of Fira in May 1980, Andreas Friedrich found fragments of pottery from ruined Bronze Age houses. The site was later excavated by the archaeologist Mariza Marthari.

When I went to see what he had, Andreas had already found another piece that fit together with the first one. At the same moment I saw more pieces that were still in the wall. When I asked him where he thought the shards had come from, Andreas looked up toward the wall and said they must have come from somewhere up there. Looking up, we could see fragments of a man-made wall and more shards. We agreed that archaeologists should be informed of this discovery and immediately went to Charalambos Sigalas, who was working at that time in the Akrotiri excavations. Together with him we examined the site more closely and within a few moments he had loosened more painted shards from the walls. Sigalas had seen enough; he reported the findings to Athens.

This site was then closed off and a year later the archaeologist Mariza Marthari from Athens excavated it and described it in scientific reports (Marthari, 1983, 1988, 1989) (Fig. 10.22.) The excavation revealed three oval rooms connected to one another and buried under 35 meters of pumice from the Minoan eruption. The inhabitants had dug into a blocky volcanic layer that can be correlated stratigraphically with the Cape Riva ignimbrite. Steps connected two rooms to each other. The third was partly destroyed by the pumice mining that had left only the remains of a wall. This room was also connected to the others by steps. The stones in the walls were taken from the surroundings and soil had been used to hold them in place. The three rooms are close to the present caldera and their floors descend like steps down toward the caldera.

Mariza Marthari was able to give an age of about 2000 years BC for the ceramics, which would place them as Early Middle Cycladic. According to archaeologists, they are the best-preserved known remains from that period (Fig. 10.23). In the evening while we were sitting on the terrace at Fira watching the sun set over Therasia, Andreas said, 'I did not know that you had such an exciting job'.

The findings on northern Thera

As our experience has shown, seemingly unimportant findings can sometimes provide unexpected, important information. While looking for blocks of limestone among the ejecta of the Minoan eruption near Megalo Vouno, Ulrike Eriksen found painted pottery sherds with pumice attached to them. At the time there was nothing sensational on northern Thera, and no archaeological findings were known from this particular place. The site was located on the inner wall of the caldera where we hoped to find evidence to aid in our reconstruction of the pre-Minoan island. It was the same area where Heiken and McCoy (1984) had noted deeply weathered rocks. They reasoned that it must have been part of the pre-Minoan caldera wall.

I found painted sherds and chips of obsidian on the under side of the Bo_I layer in the former quarry north of Oia, and the layer containing the sherds was dipping into the present caldera. A large part of the former Bronze Age surface is accessible in this same quarry. Obsidian tools found here indicate that there was a Bronze Age settlement close to the caldera wall.

Findings in the Karageorghis quarry

Until about 1982 pumice was being mined in the Karageorghis quarry located at the caldera rim north of Athinios. At that time my co-workers and I measured some profiles there and collected fossil plants. We had concentrated on this locality because we hoped to find plant remains in the Bronze Age surface that had been exposed where the pumice had been removed. There, I found a black spot that covered about one square meter of the surface. It turned out to be a pit filled with charcoal. From this pit my colleague Rud Friborg and I were able to gather five kilograms of charcoal. A sample was given to the museum in Fira and the rest was used for radiocarbon dating. Henrik Tauber of the Radiocarbon Laboratory in Copenhagen measured a radiocarbon date of 1450 BC corresponding to 1770 calendar years BC. This was in good agreement with other data from Akrotiri (Friedrich *et al.*, 1980b).

About two hundred meters to the south from the charcoal pit, I found pieces of painted pottery that were still on the Bronze Age surface (Fig. 10.24). The pieces were taken by workers of the museum of Fira who were able to restore the original vessels. Not far from this place beneath an overhanging wall workers in the quarry had discovered human bones which had fallen from the wall above. Charalambos Sigalas, the archaeologist who had investigated the site, told me that it had been a Bronze Age graveyard and that in

Figure 10.22 Some of the oldest dwellings from the Bronze Age are found south of Fira in a pumice quarry (Phtellos) close to the present caldera rim. They were built around 2000 BC. Steps connect the oval rooms with one another. The floors of the rooms are not at the same level, but descend progressively in the direction of the caldera.

Figure 10.23 A painted Cycladic jug from an excavation in the Phtellos quarry south of Fira. It is about 23 centimeters high.

the same area Christos Doumas had found a Middle Cycladic ewer (Doumas, 1983, fig. 34) from the graves.

On the caldera wall below the village of Megalochori there is a former quarry that was active till the 1950s. The mining removed only the lowermost part of the Minoan pumice series. The exposed surface gave us an exact picture of the former topography, but it has since been almost completely covered by garbage. In the autumn of 1976 I found on this surface three light 'rings' with a diameter of about 15 centimeters that were the result of pottery still *in situ*. On this place there was also a mound of 3 square meters. I reported it to the archaeologists.

When I visited this site at the beginning of September 1989 together with Doumas, the paved ground was still visible. The guard of the excavation at Akrotiri, who sometimes accompanied us, found some painted pottery sherds on the steep slope of the caldera wall. It was decided that the site should be excavated. The pottery close to the caldera wall was situated in the lower level of the paved ground. It was very close to the rim and from there one could have an excellent overview of the whole caldera.

Findings in the Mavromatis quarry

North of the town of Akrotiri there was a quarry that was still active in 1987. It is named after the owner, Mavromatis. We have often found painted sherds and the remains of walls during our geological study, which began in 1975. In 1987, the Czech geologists Antonin Paluska and Kveta Paluskowa visited me at Aarhus and informed me that while studying soil

Box 10.4 Archaeological findings in the Mavromatis quarry

'On the west side of Thera in the caldera area ranging from Fira to Akrotiri one has observed from time to time prehistoric buildings and graveyards. In this area there were for more than 100 years quarries where pumice was exploited.

The Mavromatis quarry is situated in the Akrotiri area (between Akrotiri and Megalochori). In the summer of 1979 pottery sherds of large storage vessels were found as well as walls of buildings that could not be excavated due to the lack of money. In June 1982, the activity in the quarry had expanded to this area resulting in the total destruction of these buildings. The quarry activity in this area consists of 50 000 square meters and is situated west of the St. John's Church to the area of Mavros where the prehistoric findings were made. Here one can see that these prehistoric buildings occur in the whole area. At two places, one found the remains of walls of two other buildings as well as pottery sherds that could be classified as Middle Cycladic and Late Cycladic. They show that in this area buildings had existed in a period of more than 500 years.'

After Christina Televandou (translated from Greek by Achilleas Frangopoulos)

Figure 10.24 Clay vessels found beneath the Minoan pumice in the Karageorghis quarry on Thera. Human bones and charcoal were also found in this quarry.

Figure 10.26 The ship fresco of the Akrotiri excavation is one of the most valuable illustrations of the Bronze Age. Only a part is depicted here. A possible interpretation of the landscape could be that the Pre-Kameni Island is shown on the left surrounded by concentric rings of water. From the National Museum, Athens, courtesy of Professor Doumas.

Figure 10.25 The uppermost tuff layers on the Akrotiri Peninsula fill erosional channels radiating from the center of the present-day caldera. They follow the inner slope of the caldera toward the interior, thus proving that there was a basin in that area before the Minoan eruption.

profiles in this quarry they too had observed Bronze Age ceramics. They informed the staff at the excavations at Akrotiri, and as a result the locality was later excavated by the archaeologist Christina Televandou from Athens (Box 10.4). Today, one can see ruins of houses from the Middle and Late Cycladic period close to the caldera rim.

These findings also fit into the reconstruction: the Cape Riva ignimbrite, which in this quarry dips down toward the caldera, proves that a central depression or caldera already existed at the time when the ignimbrite was formed. That means at least 21 000 years ago, as measured by radiocarbon technique (Eriksen *et al.*, 1990). Also, the layer on which the Cycladic houses were built dips into the present caldera. Thus, we get the impression that they were formerly situated close to the caldera rim, and it was presumably possible to see the water-filled caldera from that place.

However, there were more findings in support of the reconstruction: old erosional channels dated by archaeological remains (Fig. 10.25).

In the spring of 1981 I visited the area in the vicinity of the lighthouse on Akrotiri peninsula. Some Bronze Age objects had been found in this area on the surface beneath largely eroded pumice (Marinatos, 1976). In a nearby erosional channel named Kaminia the owner of the property had found a Cycladic vessel with the remains of a new-born baby, and in this same area one finds even today painted pottery and arrow heads made of obsidian (Aston and Hardy, 1990).

This Bronze Age site is in an erosional channel that predates the Minoan eruption. The channel radiates from the center of the caldera today just as it did before the this eruption. It drained water from the Akrotiri peninsula into the caldera. It was filled by the pumice of the Minoan eruption and in the course of time was restored by erosion. This site was a further argument for the existence of a water-filled caldera.

Other observations and ideas

Two American geologists, Grant Heiken and Floyd W. McCoy, presented a completely new and interesting idea at the Thera conference of 1989. In 1984, they had published a reconstruction of the Bronze Age caldera using diagrams showing the island from different viewpoints. With these models they attempted to show that the hilly landscape on the miniature fresco at Akrotiri is identical to a natural perspective of the former landscape. In their opinion one can see on the fresco the water-filled caldera with Profitis Elias in the background. Their comparison was quite convincing, and if true it offers a further clue for the understanding of the frescoes at Akrotiri.

On the day they presented their arguments to the congress there were reactions from several sides: if this fresco depicts the caldera it also shows the Pre-Kameni island as shown in our reconstruction (Fig. 10.26).

In the meantime, Druitt and Francaviglia (1992) presented the results of studies in which they showed their earlier observations on the inner side of the caldera and supported these by further measurements. Of special interest is the finding that at various places the Minoan pumice was 'plastered' on the inner side of the caldera. Their reconstruction of the island before the Minoan eruption is in good agreement with ours (Friedrich *et al.*, 1988 and Eriksen *et al.*, 1990).

11

Geological observations and the legend of Atlantis

Several important questions are answered by the new geological interpretation of the appearance of Santorini before the Minoan eruption, but many still remain. Perhaps the most tantalizing mystery is Plato's account of Atlantis and the flourishing culture that suddenly disappeared into the depths of the sea.

Was Atlantis situated on the Bronze Age island?

When the sensational frescoes were found at the end of the 1860s, interest in Plato's report of the disappearance of Atlantis took on new life. Plato (427–347 BC) was one of the great philosophers of the ancient world and was the friend and disciple of Socrates. The island of Atlantis that he described in his dialogs 'Critias and Timaeus' (see Appendix 1) bears many similarities to the pre-Minoan Santorini. He tells of the disappearance of a round island consisting of concentric rings and populated by people of high culture. Plato tells us how Critias got the story from his grandfather who also had the same name. The tradesman Solon (640–560 BC) is said to have brought the story back from his travels. Critias' grandfather had heard the story from Dropides who was a friend and relative of Solon, 'the greatest of the seven wise men'. On his travels to Egypt, Solon had visited the town of Sais on the Nile delta, where he got the story of the disappearance of Atlantis from Egyptian priests, who had a written version of it. Critias says: 'this written report was in the hands of my grandfather (Critias) and now it is in my hands and I studied it thoroughly when I was a child'. Plato learned of this report about 300 years after Solon traveled to Egypt, and he passed it on in the form of dialogs. Since that time, it remains the source of endless speculations, presumptions, and hypotheses. In the dialogs, Plato states repeatedly that this story was not invented but was based on fact. But even in classical times, just as now, some of the scholars interpreted this as a fantasy. One of these was Aristotle (384–322 BC), another great philosopher of antiquity and student of Plato. He took the story of Atlantis as an invention of his teacher: 'the man who dreamed up this island made it disappear again'. Others, however, took the story literally. One of these was Proclus (AD 410–485), one of the later commentator of Plato's work. He wrote in his Comments on Plato's Timaeus (76.1 to 10)

the discussion about the people of Atlantis where (some say) . . . that it is a true and ancient story to belong to Crantor the first commentator of the work of Plato. Furthermore, according to Crantor, the account of Plato . . . that says that he had invented the story . . . is not an invention but he had copied it from the Egyptian institution. As a proof Plato (23.A-4) refers to the Egyptian priest who said that those items are chiseled into the columns and preserved till the present day.

Crantor lived in the third century BC; his original work is lost.

This is in contrast to the Greek philosopher Posidonius (135–51 BC), who later proposed that the story of Atlantis was a mixture of reality and fantasy. The geographer Strabo (67 BC–AD 23) reports about Atlantis in his work *Geographica* (2.3.6–7). He says

Figure 11.0 Part of the 'Crocus gatherers fresco' of Akrotiri shows beautifully dressed girls with golden jewelry on their head, ears and arms. Their attire suggests that they are performing a ceremonial ritual. From the National Museum, Athens, courtesy of Professor Doumas.

with regard to Posidonius: 'concerning that point he is right to cite Plato who says that the story of Atlantis is not a fiction'. And we learn from Ammianus Marcellinus (AD 330–400) that at Alexandria the legend of Atlantis was considered a historical fact.

Today, the story of Atlantis is the topic of endless discussion. So much has been written that a brief overview cannot possibly give an adequate account in a book such as this. A story that the press treats like reports of Yetis in the Himalayas cannot, as many say, be taken seriously. Some would like to make the fairy tale of Atlantis off limits and avoid wild conjectures. While no one likes to be called a utopian dreamer, it may be worth looking at the core of the Atlantis legend as it is seen by a number of modern scholars. In my opinion, no one can describe the natural history of Santorini without considering the legend that recounts the fate of a Bronze Age island that suddenly disappeared into the sea.

The Minoan culture and Atlantis

As far as one can tell from the literature, the French scholar Louis Figuier (1872) was the first to connect Santorini with Plato's report about Atlantis. At that time, the French archaeologists Mamet and Gorceix (1870) had made sensational findings under the thick pumice layer at Akrotiri, and Fouqué (1869) had just published his work entitled *A Prehistoric Pompeii*. After this there was a period of several decades in which little was said about Atlantis in relation to Santorini.

In the twentieth century, however, a few scientists revived the discussion. One of these was Frost, who wrote an anonymous article in *The Times* (19 January 1909) about a possible connection of Atlantis with Crete, where Sir Arthur Evans had just uncovered the Palace of Knossos. Frost argued his case in more detail in a later article in *The Critias and Minoan Crete*, where he also acknowledges his authorship of the article in *The Times*. He had drawn special attention to the last part of Plato's text, where he deals with the bull culture of the Atlantis people. He pointed out that this was also part of the Minoan culture (*Critias* 119C–120D):

In the sacred precincts of Poseidon there were bulls at large; and the ten princes being alone by themselves, after praying to the God that they might capture a victim well-pleasing unto him, hunted after the bulls with staves and nooses, but with no weapons of iron; and whatever bull they captured they led up to the pillar and cut its throat over the top of the pillar, raining down blood on the inscription

The theory was given a new impulse when, just before the Second World War, the Greek archaeologist Spyridon Marinatos (1939) excavated a number of places on Crete including the Minoan villa at Amnissos, a harbor of Knossos. He saw there a connection between the decline of the Minoan culture and the eruption of the volcano of Santorini. In 1950, he published another article with the title 'On the legend of Atlantis'. In a later version of the same article (Marinatos, 1972), he emphasized his views on Santorini and Atlantis: 'I continue to believe as I described many years ago in the present paper, that the eruption of Thera may be the raison d'être to the Atlantis literature.' In yet another work he said (as quoted by Galanopoulos, 1981): 'The only myth that could be interpreted by the Thera explosion is the tradition for Atlantis by Plato . . . It seems that the disaster of the small Thera has been exaggerated in the existence of the huge mythical Atlantis.'

In his 1972 work Marinatos confirmed his opinion that the legend of Atlantis had a historical core, pointing to old Greek traditions which were known to the people living on the Cyclades more than one thousand years after the Minoan eruption. Pindar, who lived around 522–441 BC, reported in his Odes (*Paean*, IV.40–45), nearly 100 years before Plato and much clearer than him, about the same tradition where Zeus and Poseidon made it disappear into the depths of the sea:

Know ye that I fear war with Zeus, I fear the loudly thundering Shaker of the earth. They, on a day, with thunder-

bolt and trident, sent the land and a countless host into the depths of Tartarus, while they left alone my mother and all her well-walled home.

On the basis of further texts by Pindar, Marinatos emphasized that this area was situated in the Cyclades, a group of islands to which Santorini belongs.

Something resembling the Minoan eruption is found in Homer's *Odyssey*, an important classic written in the eighth century BC. When the nautical pilots from Phaiaken had led Odysseus back to Ithaca and the ship was changed to stone, their king Albinos called out:

Alas, my people, now are fulfilled the antique prophecies my father used to tell me, how Poseidon was so contraried by our granting free passage to all and sundry that upon a time he would destroy one of our best ships as she came in through the ocean-haze to land: and then would obscure our city within a wall of hills.

From *The Odyssey of Homer*, Book 13, p. 186. Translated by T. E. Lawrence, Oxford University Press 1991.

Furthermore, it is worth mentioning that the father of Alcinous, Nausitoos, was the founder of the Phaeacian City named Scheria and that he was the son of the 'shaker of the earth, Poseidon'. It is reported that the city of Scheria, which the angry Poseidon had covered with masses of stone, was within a day's sailing of the island of Euboia. This is also the distance to Santorini.

The Atlantis theory of Galanopoulos

In my opinion, the most fascinating argument for connecting Santorini to the legend of Atlantis was that of the Greek geophysicist, Galanopoulos; he took a new approach to the problem by considering geological evidence (Box 11.1). In 1957 he reported the results of his studies in a quarry at Fira ten years before Marinatos began the excavations at Akrotiri. In 1960 he related the Bronze Age findings on Santorini to the island of Atlantis, and, together with Edward Bacon, the archaeological editor of *Illustrated London News*, he elaborated his theory in the book *Atlantis, The Truth Behind the Legend* (1969). The authors were able to offer convincing scientific arguments for connecting Santorini with the Metropolis of Atlantis. More will be said about this book later.

Since then, several other persons have discussed the Minoan culture in the context of Atlantis. The most notable of these studies were those of the American oceanographer James W. Mavor in his *Journey to Atlantis* (1980), the British historian John V. Luce in his book *The End of Atlantis: New Light on an Old Legend* (1973), and the Greek archaeologist Nicholas Platon *Zakros, The Discovery of a Lost Palace of Ancient Crete* (1971). The present director of the excavations of the settlement at Akrotiri, Christos G. Doumas (1983), has also said a few words on this theme in his book, *Thera, Pompeii of the Ancient Aegean*.

The archaeologist Nicholas Platon (1971) and the historian John Luce (1973) give overviews of the Greek and Egyptian roots of the Atlantis legend. They interpret them in connection with the excavations on Crete and other places in the Mediterranean. In their view, the country the Egyptians called '*Keftiu*' is equivalent to Crete, an island with which they carried on an extensive trade (Fig. 11.1). According to Platon, Luce, and others, Keftiu could be the same as the land of Kaphtor of the Bible. We read in Amos, 9.7: 'Have not I brought up Israel out of the Land of Egypt? And the Philistines from Kaphtor, and the Syrians from Kir?'. In the text of Jeremias (47.4,): 'For the LORD is destroying the Philistines, the remnant of the coastland of Kaphtor.'

There are also Semitic sources dealing with an island Kaptara that presumably is the same as Kaphtor. Helck (1979) gives several examples of the connection of Egypt and Asia Minor with the Aegean as early as the seventh century BC. The name Keftiu is often mentioned in these accounts.

Where could Keftiu have been situated? Strange (1980) offers arguments for connecting Keftiu with

Box 11.1 The Atlantis theory of Galanopoulos and Bacon

These authors interpret Plato's story as a historical account rather than a legend or myth. They argue that this event occurred in the Bronze Age between 2100 and 2200 BC and at two places, a small round island with a radius of 9.5 kilometers (the Metropolis) and a much larger rectangular island. They dismiss Atlantis theories that offer no geological reason for the sudden disappearance of the island. They find that the only logical scene of those events is the eastern Mediterranean and that the usual identifications of the pillars of Hercules mentioned by Plato with the Straits of Gibraltar are wrong. Instead, they believe that the name was applied to the strait between Malea and Cape Matapan on Peloponnese.

They can also show that there was volcanic activity in the eastern Mediterranean during the Bronze Age and at the same time a small inhabited island at the center of Santorini disappeared. They therefore consider the identification of 'Stronghyle–Santorini' with the Metropolis as very probable. While conceding that Plato never said that the Metropolis was located on a volcano, they point out that the volcano had passed through a long period of quiet. In support of this, they point to the very fertile soil that was said to surround the acropolis and remind us that such fertile soils develop on volcanic ash. The colors of the rocks mentioned by Plato – red, black, and white – are commonly seen in volcanic areas and are especially notable on the island of Thera. Even the warm and cold springs mentioned by Plato fit a volcanic setting, because they are also found on Santorini.

Plato's description of the form and structure of the metropolis with its central cone agrees, according to Galanapoulos and Bacon (1969), with the central cone of Stronghyle–Santorini, since the measurements of the Metropolis are within the same range of size. Their illustration is shown in Fig. B.11.1. They believe they can even see in the long-accepted model of Santorini based on the British Admiralty chart of 1916, traces of the moats of the Metropolis. The central harbor is said to have been in the circular areas between Nea Kameni and Thera and between Nea and Palea Kameni. In their sketch they place the outline of the Metropolis of Atlantis within Santorini. In their opinion, if one compares this sketch with the model of Santorini one can see traces of the channels and the bottom of the caldera. The latter would be the zone of water described by Plato. Furthermore, the distance from the central cone of the volcano would be equivalent to the width of the corresponding water to the hill on which the temple of Poseidon is situated. They observe a discrepancy however. In Plato's *Critias* (113 C) the acropolis has a distance of 50 stadia from the sea, while in *Critias* (117 E) the outermost zone that surrounded the acropolis was situated 50 stadia from the sea. For their drawings they used the values given in the second text, but how do we know the first text would not be more appropriate? Then the radius of the Metropolis should be two kilometers less than shown in the drawing and this would be almost exactly the same radius as Santorini has today. Moreover, the gorge between Thera and Therasia would be exactly as long as the channel that connected the sea with the inner moat of the old metropolis.

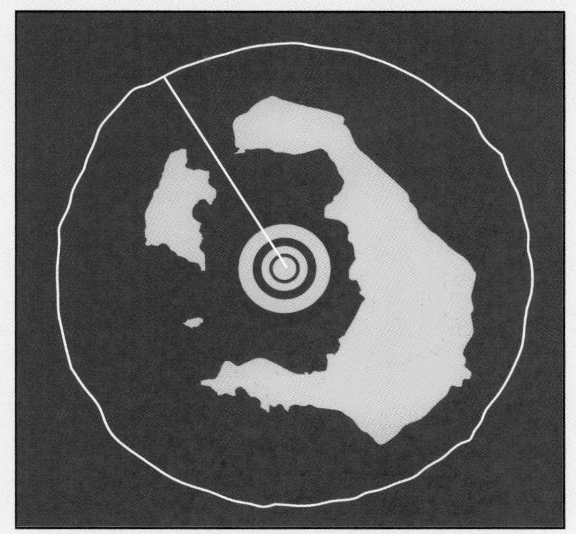

Figure B.11.1 An illustration from the book *Atlantis* by Galanopoulos and Bacon (1969) shows a comparison of the size of the Metropolis of Atlantis described in Plato's dialogue *Critias* (rings) with the present shape of Santorini.

The authors think that this coincidence is noteworthy but even more remarkable is the form of the seaward end of the under-sea gorge. According to Plato, new inhabitants of Atlantis had widened this mouth in order for the largest ship of that age to enter the moat. Galanopoulos and Bacon admit that the supposed remains of the moat on the bottom of the caldera could be natural formations, but they find it difficult to believe that the opening of the underwater gorge between Thera and Therasia could have originated by chance since its depth and form exclude its formation by erosion. They consider this fact together with the length of the underwater gorge and the connecting channels of Atlantis as strong arguments that these trenches could not be accidental. If one follows the description of Plato then it would be obvious that the ring-shaped zones of seawater should be natural channels that surround the central cones. On the other hand, the channels that connected the water belts may have been man-made, at least in part. According to Plato they were made by Poseidon, the god of earthquakes, before any ships existed (*Critias* 11C-E). They contained sea water and were therefore connected to the sea. It is unlikely that this connection was suited for the passage of ships, since, as Plato says, the descendants of Poseidon had to widen this channel (*Critias* 11C-E) in order that ships could pass through it. They even constructed a roof over it and created a tunnel.

Galanopoulos and Bacon also offer an explanation of how the people in the second millennium BC could solve such an immense problem. They find two similar projects that were completed at that time. One is the Megalithic construction at Carnac in France; another is the one-kilometer long tunnel at Eupalinos on Samos that was erected as an aqueduct by the tyrant Polycrates in the middle of the sixth century BC. According to Galanopoulos, Bacon and others, the latter of these examples is in the capital of Plato's Atlantis.

Cyprus. Others, however, disagree. Among the latter, an archaeologist working on Crete, Erik Hallager (1988), is of the opinion that Keftiu is a synonym for Crete.

According to Strange, the clue to the geographical position of this country is given in an inscription found in the tomb of Rekmire in Thebes: 'Keftiu and the island in the middle of the sea.' The text is accompanied by illustrations of bearers bringing gifts characteristic of the Aegean region. This tomb dates from the 18th dynasty at the time of Pharoah Tuthmosis III (1490–1436 BC).

Luce explained the disappearance of Keftiu in the following text:

While Tuthmosis III was campaigning in Canaan, and using ports frequented by Minoan traders, a great and terrible disaster overwhelmed Keftiu and its island dependencies. The Egyptians must have heard about the cataclysm from refugees and sailors. They probably recorded something about the physical disappearance of a large part of Thera, and the shrouding of Crete in darkness and dust.

Luce (1969)

More recent archaeological findings on Crete indicate that the damage by the eruption was much less than Luce and other researchers had thought only a couple of decades ago. Contact between Crete and Egypt was not interrupted by the volcanic event on Santorini. This is clearly seen from an Egyptian inscription from the time of Amenophis III (1403–1364 BC) in which the name of Keftiu is mentioned along with Knossos and Amnissos on Crete.

In the image-rich writing of the Egyptians one frequently finds the word 'pillar' in connection with Keftiu. This led Luce to connect this land with Crete: '. . . it was an island with a pillar that held up the sky'. This picture, however, can be interpreted in a different way. During a volcanic eruption, an eruption column is formed which literally ascends up to the heavens and in this way creates a connection between heaven and earth. Thus it could be argued that the Egyptian term denotes a connection to Santorini volcano.

Luce offers a good answer to the question of where Atlantis was located. According to Andrew (in Luce, 1969), Plato interpreted the notes of Solon

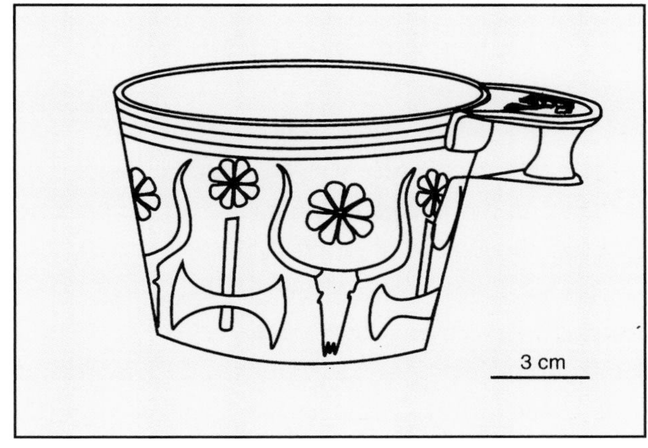

Figure 11.1 In the tomb of Theben in Egypt several recent discoveries, such as those of Rekmire, Useamon and Senmut (Senenmut), indicate contact between Egypt and the people from Keftiu. This illustration from the tomb of Senmut shows gift bearers bringing Vapheio cups with ornamental bullheads that, according to some researchers, are from Crete (after Davis, 1930). A silver cup from the Greek locality at Midea is shown for comparison. (After Davis, 1977, Fig. 211.)

incorrectly: instead of 'greater than Libya and Asia' as in Plato's text, one should read 'between Libya and Asia' (since the Greek words in question, *meson* and *mezon*, differ in only one character).

Galanopoulos and Bacon (1969) have also considered the timing of the disappearance of Atlantis. Plato says that this happened 9000 years before the story was written, but this cannot be correct. For this they have a better explanation: since the story was brought by Solon from Egypt and later was translated into Greek it is possible that an error was introduced in translating the measurement. Because the Egyptians used a different counting system, it should read nine hundred instead of nine thousand. The age would then fit perfectly according to Galanopoulos and Bacon: Plato lived about 300 BC, and Solon traveled to Egypt 300 years earlier. Adding 300, 300, and 900, one gets 1500. This means that Atlantis disappeared around 1500 BC, a time that agrees fairly well with that of the Minoan eruption.

Galanopoulos writes:

Nevertheless, one might venture to say that there is now convincing evidence at least for the origin of the story. It is now generally accepted that Thera's frescoes reveal the sophistication of a sea-trading people and the source of the legend of Atlantis. Even if Plato made up his story, as Aristotle and his adherents maintain, the story is true. Beyond any doubt there was in the Aegean Archipelago an advanced civilization in the Bronze Age destroyed by 'tremendous earthquakes and overwhelming floods'.

The oceanologist James Mavor tried to connect Atlantis with Santorini, but his results were ambiguous (see Box 11.2).

Modern geology and Atlantis

The arguments brought forward by Galanopoulos and Bacon are convincing, especially those that connect the Metropolis with a volcano. Less convincing, however, is their opinion that it is possible to see the moats of the Metropolis in a structural model of the caldera. As we see it now, the Pre-Kameni island, which was situated in the middle of the caldera, was entirely destroyed by the Minoan eruption. As a result of the eruption the site of the former island became the deep northern basin of the caldera (Fig. 11.2). The topography of the caldera continued to change as the caldera walls, both above and below sea level, slumped into the newly deepened basin. The revival of volcanic activity after 197 BC added to these changes.

The structure of the caldera today has been revealed by geophysical investigation (Figs. 11.2 and 11.3). Perissoratis and his coworkers (1995) delineated the northern basin between Nea Kameni and the northern wall of the caldera. The bottom at 390 meters below sea level is covered with a layer of loose sediments up to 120 meters thick. The situation in the southern part of the caldera is quite different. South of the Kameni Islands the caldera floor is only about 290 meters deep, and the bottom sediments are only about 30 meters thick. In a few places, the flat-lying beds of post-Minoan sediments are domed up by intrusions.

Galanopoulos and Bacon cite the red, white, and black rocks and the cool and warm springs mentioned by Plato, as evidence that Atlantis was a volcanic island, but these features are found on other volcanic islands as well as Santorini. The shape of the round island that disappeared, however, fits only Santorini.

One could add to the distinctive features of the island pointed out by Galanopoulos and Bacon the mineral deposits that occur on the island that disappeared. With the exception of gold, many metals are found on Santorini, and they were presumably accessible to the Bronze Age people. Mineral deposits of copper, lead, zinc, talc, and silver may have been exposed and were probably exploited during the Bronze Age. Today, the Minoan pumice covers a large part of the ground where these minerals occur. There is no way of knowing whether there was gold on the island, but it has not been identified by any of the many geologists who have examined the island. If gold had been found on Santorini, it would have been easier to explain the apparent wealth of the settlement at Akrotiri. They had gold at that time and also ivory, for we see the beautifully dressed girls on the frescoes wearing them (Fig. 11.0). Even the form of the island, which according to Plato was surrounded by high mountains, is reflected in the caldera walls. However, the most important argument of the theory of Galanopoulos and Bacon is the similarity of the concentric rings created by Poseidon to a common structural feature of calderas.

As we have seen in the preceding chapter there is overwhelming evidence for a flooded caldera with an island in the middle before the Minoan eruption (Fig. 11.4C). This was unknown to Galanopoulos and Bacon. We also know that the interior island was present at least as early as the Cape Riva eruption of 21 000 years BC (Fig. 11.4B) and was still present at the time of the Minoan eruption (1640 BC). The radiocarbon data for the marine fossils in stromatolites and snails, which were living in this time span, give us an

Box 11.2 The voyage to Atlantis

The oceanographer James Mavor tried to use the modern tools of marine geology to prove the theory of Galanapoulos and Bacon. He and Galanopoulos investigated the sea bottom of the caldera in 1965 with the research vessel *Chain* and looked for traces of Atlantis in the caldera and on the outer side of Santorini. As a designer of the underwater vessel *Alvin*, Mavor was able to collect a group of competent scientists around him. Their undoubtedly important investigation, however, was misinterpreted in the world press.

On the 10th of September the *New York Times* published under the title

MOAT BELIEVED TO BE PART OF ATLANTIS IS FOUND IN AEGEAN SEA

Prof. Angelos Galanopoulos announced today the discovery of 'most convincing proof' that the legendary city of Atlantis had been found in the Aegean Sea. The professor, who is a leading Greek seismologist, said that the outline of a wide moat had been detected 1300 feet underwater in the submerged part of Thera island.

A Greek–American team of scientists led by Dr. James W. Mavor of the Woods Hole Oceanographic Institution in Massachusetts has been working for a week in Thera to explore the evidence of Professor Galanopoulos' theory that Atlantis should be identified with the Cretan empire that existed in the Aegean in 1500 BC.

Prof. Galanopoulos said the moat was probably part of Atlantis' sacred island, the metropolis.

An announcement from the seismological laboratory of Athens University, which is headed by Professor Galanopoulos, said that the discovery had been made by the research vessel Chain, which belongs to the Woods Hole institution.

The Chain made a seismic profile of the Aegean island of Thera, two-thirds of which was sunk 1300 feet below sea level by a volcanic eruption around 1500 BC. Electronic devices aboard the research vessel traced the moat under a thick layer of volcanic ash.

On September 10, the *New York Times* reported his more detailed comments under the headline ATLANTIS EXPLORERS RETURN TO ATHENS.
The story stated:

Dr. James Mavor of the Woods Hole Oceanographical Institution in Massachusetts, who headed the team, said today: 'There are shreds of evidence which, if put together, point toward a confirmation of the theory that the lost continent should be identified with the Minoan Empire, which ruled the Aegean archipelago and Crete about 1500 BC.'

The team was working on a theory by leading Greek seismologist, Prof. Angelos Galanopoulos who believes Atlantis was a commonwealth of islands led by Crete that was destroyed by gigantic volcanic eruptions between 1520 and 1420 BC.

With this background it is quite understandable that Mavor's planned expedition to Santorini the following year did not take place under the best conditions. His exciting reports on the *Voyage to Atlantis* (1980) provides an excellent example of what can happen when an enthusiastic American scientist does not have the tough personality needed to work in foreign countries. The complications, difficulties, and misunderstandings between Mavor and the excavator of Akrotiri, the Greek archaeologist Marinatos, could be anticipated from the articles in the newspapers. It is interesting to see this story from the Greek perspective. For this purpose one should read the book by Lois Knidlberger (1975) entitled *Santorini: Island Between Dream and Day*, in which the facts and theories about Atlantis are discussed. Knidlberger is obviously on the Greek side in the dispute between Marinatos and Mavor.

Figure 11.2 (Opposite) The present caldera of Santorini consists of four basins. The northern one reaches a depth of 390 meters below sea level, while the other three are only 290 meters deep. The Kameni islands are situated in a southwest–northeast trending zone of weakness. A second zone with about the same orientation but displaced slightly to the north extends from Megalo Vouno to Kolumbo volcano outside the caldera northeast of Thera. Contour lines are 20 m apart. From Perissoratis (1995).

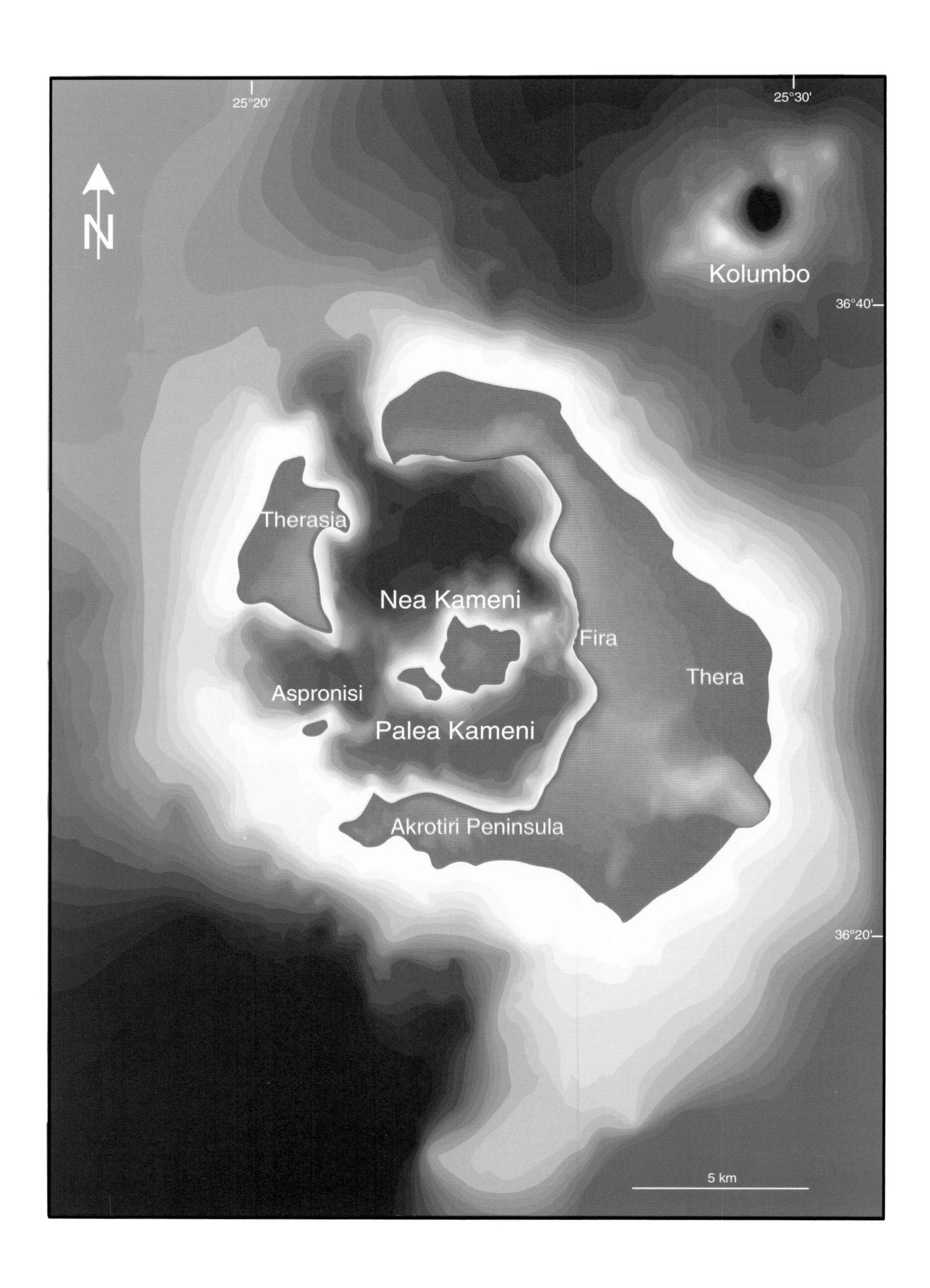
25°20'
25°30'
N
Kolumbo
36°40'
Therasia
Nea Kameni
Fira
Thera
Aspronisi
Palea Kameni
Akrotiri Peninsula
36°20'
5 km

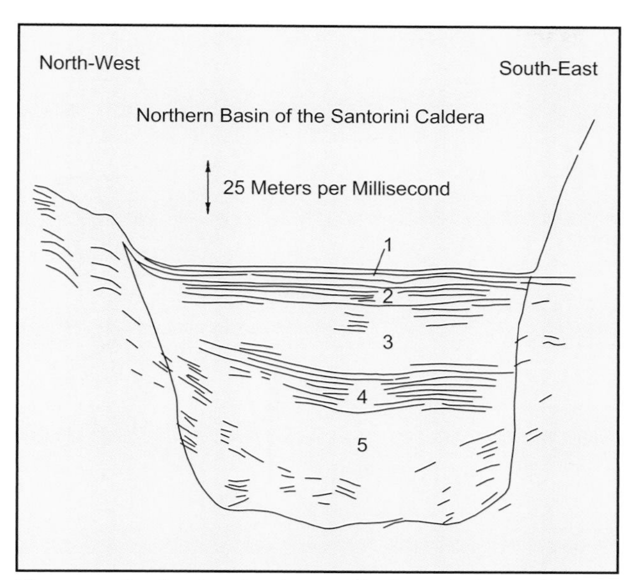

Figure 11.3 A seismic reflection profile through the northern basin of Santorini caldera shows five units made up of nearly horizontal layers of loose sediments. They have a total thickness of about 120 meters. (After Perissoratis, 1995.)

age of 13 000 to 10 000 years BC. That means that several thousand years before the Minoan eruption the Riva caldera already existed and was filled with sea water. Several observations enable us to deduce that a long period of quiet preceded the Minoan eruption. On Thera one can see places where the Cape Riva ignimbrite is directly overlain by deposits of the Minoan eruption. At Akrotiri, for instance, houses were built directly on the ignimbrite. Moreover, the deposits under the Minoan eruption are deeply weathered. The only event we can trace to this period is the formation of the Pre-Kameni island within the caldera. The southern basin of the caldera already existed at that time, while in the north there was a shallow part of the caldera floor (Fig. 11.4A to C) where the stromatolites grew until they were disrupted by the Minoan eruption. The form of this caldera had certain resemblance to the one drawn by Galanopoulos (Fig. 11.4C and D) (Galanopoulos and Bacon, 1969).

Nested calderas are formed (Fig. 11.5) when, during a strong eruption, a large part of the magma chamber is quickly emptied and the support for the roof over the chamber is removed. When the volcanic edifice collapses into the magma chamber, a funnel-shaped depression is formed. In some cases the collapse takes place on concentric ring faults that converge downward. If this fault system is close to sea level and partly flooded the resulting set of rings separated by water and rock could resemble the ring islands described by Plato (Fig. 11.4 D). An example in which ring faults are well developed can be seen at the summit of Kilimanjaro in East Africa. In this case, the concentric rings are not separated by water but by snow (Fig. 11.6). A better example of a concentric ring structure can hardly be found.

Within the limits of our present geological knowledge, it is possible that there could have been such a ring-shaped caldera system before the Minoan eruption. The other geological arguments of the theory of Galanopoulos and Bacon are equally plausible. Even their description of the disappearance of the island of Atlantis is similar to the final phase of the Minoan eruption. The 'earthquakes and floods' Plato speaks of must also have accompanied the Minoan eruption. We read in Plato's account that 'after this event the sea was inaccessible for a long period owing to the mud that the sunken island left behind.' We have only to read the word 'mud' as 'pumice' that was floating on the sea to see that this detail fits what must have occurred following the eruption of 1640 BC.

Nevertheless, these arguments still do not give us a final link between the Metropolis of Atlantis and Santorini. The theory of Galanopoulos and Bacon is fascinating, but many unanswered questions remain. Will we find the answers in the Akrotiri excavation or is it buried in the inaccessible depths of the caldera? Will answers emerge from the new excavations on the Nile delta (Bietak and Marinatos 1993) where the Hyksos palace at Ezbet Helmi (Avaris) is presently excavated? Or will the words of Zeus and Poseidon remain the only answer?

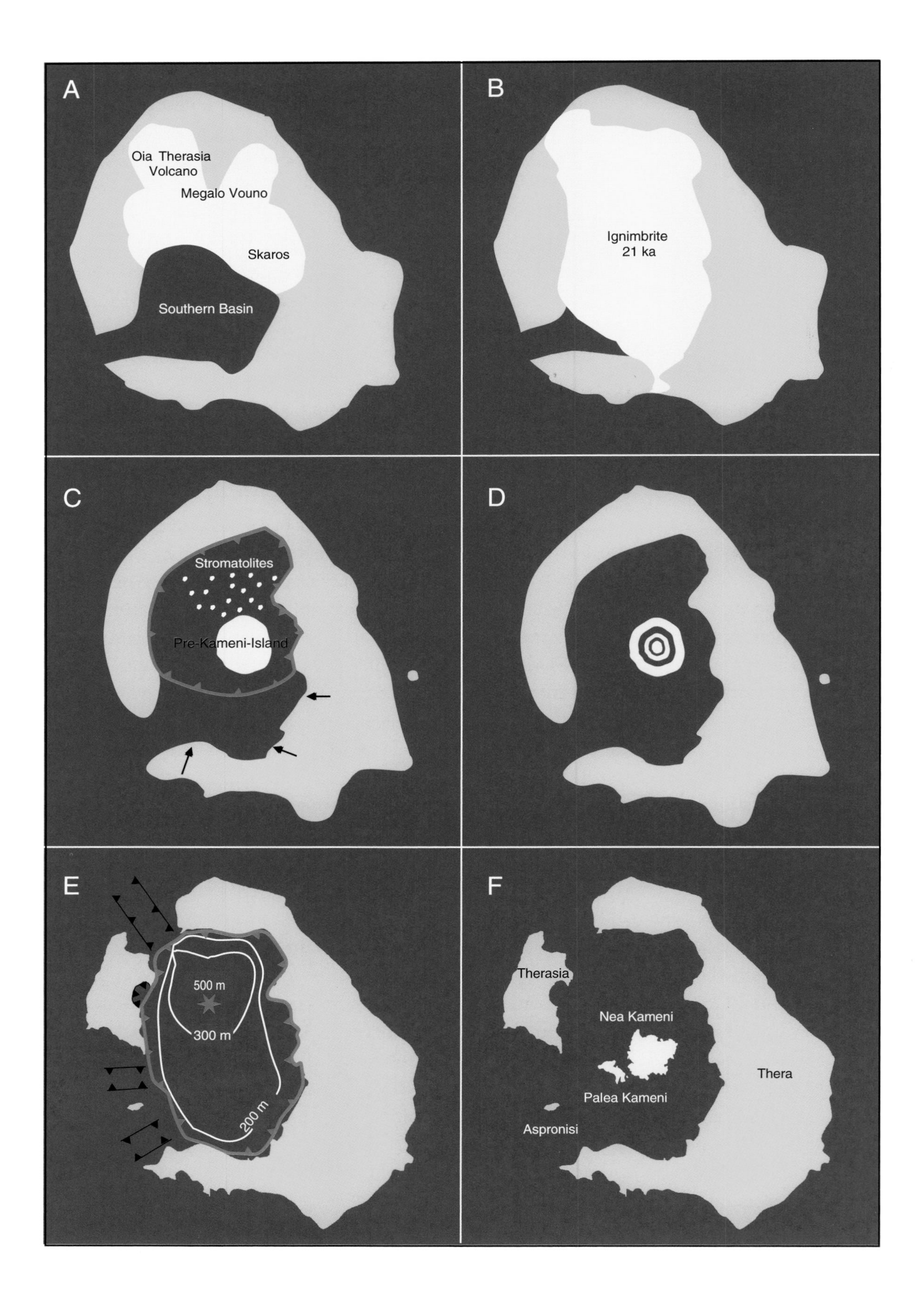
A
Oia Therasia
Volcano
Megalo Vouno
Skaros
Southern Basin
B
Ignimbrite
21 ka
C
Stromatolites
Pre-Kameni-Island
D
E
500 m
300 m
200 m
F
Therasia
Nea Kameni
Palea Kameni
Thera
Aspronisi

Figure 11.4 (Opposite) **A** In the northern part of Santorini several volcanoes were active at various times. The principal vents were at Megalo Vouno, Skaros, and Oia-Therasia volcano (Peristeria volcano) as well as Micro Profitis Elias. In the northern part of the island volcanoes formed an overlapping complex while in the south they grew in a depression that was probably filled with water.
B About 23 000 years ago an enormous ignimbrite eruption occurred in the northern part of Santorini near Cape Riva. The deposits are found in Millo Bay on Therasia, in the quarries at Oia, around the Mavromatos quarry at Akrotiri, and under the Bronze Age settlements at Akrotiri and Potamos. The age of this eruption was obtained by radiocarbon dating of trees that were buried by the ignimbrite. The characteristic rocks of the ignimbrite are also found in deposits from the Minoan eruption.
C In the north a shallow lagoon was formed where stromatolites were growing some 15 to 12 thousand years ago. Renewed volcanic activity created the Pre-Kameni island in the middle of the Bronze Age caldera. In the south a system of post-ignimbrite erosional channels draining into the southern basin indicates that this basin was deepened after the Cape Riva eruption.
D Collapse of the roof of the magma chamber located under the Pre-Kameni island led to a hypothetical ring-like structure of this island. This is also where the Minoan eruption started with a plinian phase on the Pre-Kameni island. During this and the following phases of the eruption this island was totally destroyed and its rocks were ejected together with the masses of pumice of the Minoan eruption. A water-filled funnel was produced.
E The Minoan eruption caused several changes to the ring-island: the original volcanic edifice was separated into Thera, Therasia, and Aspronisi. With time, concentric and radial slumping on the steep caldera walls gave the caldera its present form. The northern basin was formed, with an original depth of more than 500 meters. Today it is only 390 meters below sea level. In the eastern part, the products of the Minoan eruption considerably enlarged the island of Thera and the island of Monolithos was connected to Thera.
F After 197 BC, new volcanic rocks began to fill the caldera again. The islands of Palea Kameni and Nea Kameni were formed. The latter is the only active volcano on Santorini at this time.

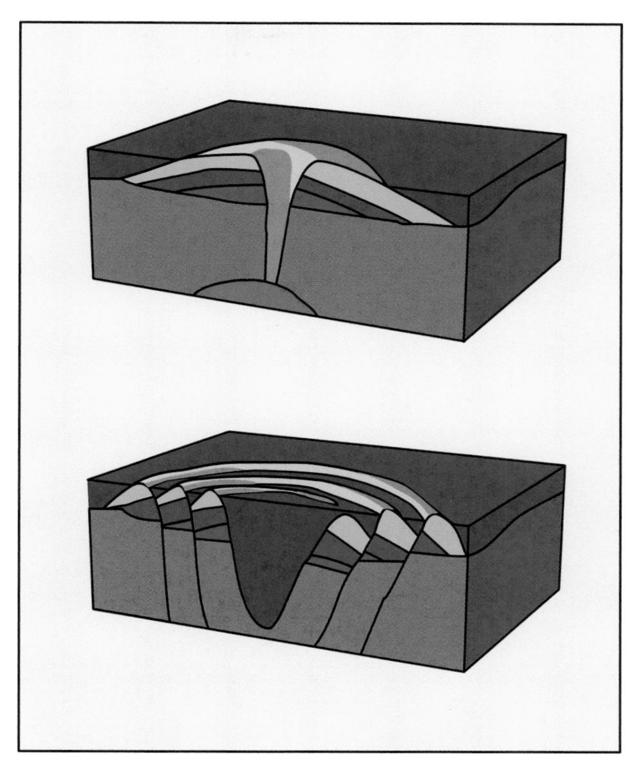

Figure 11.5 A schematic section through a volcano before and after formation of a caldera. After a large explosive eruption the upper parts of the volcano collapse into the space evacuated by the magma and create a funnel-shaped depression with a system of concentric ring-faults.

Figure 11.6 (Below) On the African volcano Kilimanjaro one can see the concentric ring system that was probably formed after an explosive eruption that had drained the magma chamber and allowed the roof to drop on concentric faults. The result is a series of concentric ridges. Photo by Maurice Krafft.

Part 4

The island is changing its appearance

12

The caldera is filling again

After the catastrophic Minoan eruption, the eruptive activity of Santorini was by no means over. For the last two thousand years, the enormous caldera that had been enlarged and deepened by the eruption of 1640 BC has been the scene of renewed volcanism: the caldera is filling again and the volcanic edifice is changing its shape.

Volcanic activity in historical time

According to an ancient proverb, 'times are changing and we are too'. This is especially true of Santorini. The Kameni islands are an example of the ever-changing dynamic earth. The southern Aegean volcanic arc to which Santorini and other islands belong is seismically and volcanically active.

Santorini, Milos, and Nisyros are all situated on the volcanic arc and can revive at any time. Strabo tells us that the island of Methana was active in 282 BC, and the island of Nisyros erupted in AD 1422. In more recent times, the Kameni islands have been the only active part of this arc. Their eruptions are related to a northwest-trending zone of weakness in the earth's crust where the magma finds a way to the surface. In addition to the Kamenis, we find in this zone the Christiana islands, situated about twenty kilometers southwest of Santorini, and another eruptive center outside the caldera about seven kilometers northeast of Cape Kolumbo (Fig. 12.1).

We have numerous reports of eyewitnesses who described the eruptions that formed the Kameni islands during the last two thousand years.

The present day caldera

The Minoan eruption of 1640 BC deepened and widened an existing caldera, but soon after the eruption was over slumping of the caldera walls began to deposit material on the floor, just as it did after the

Figure 12.0 The two Kameni islands, Nea and Palea Kameni, are situated near the center of the caldera. They were formed after the Minoan eruption and developed by repeated eruptions during the last 2000 years. The photograph shows a crater on Nea Kameni and, in the background, the caldera wall of Thera.

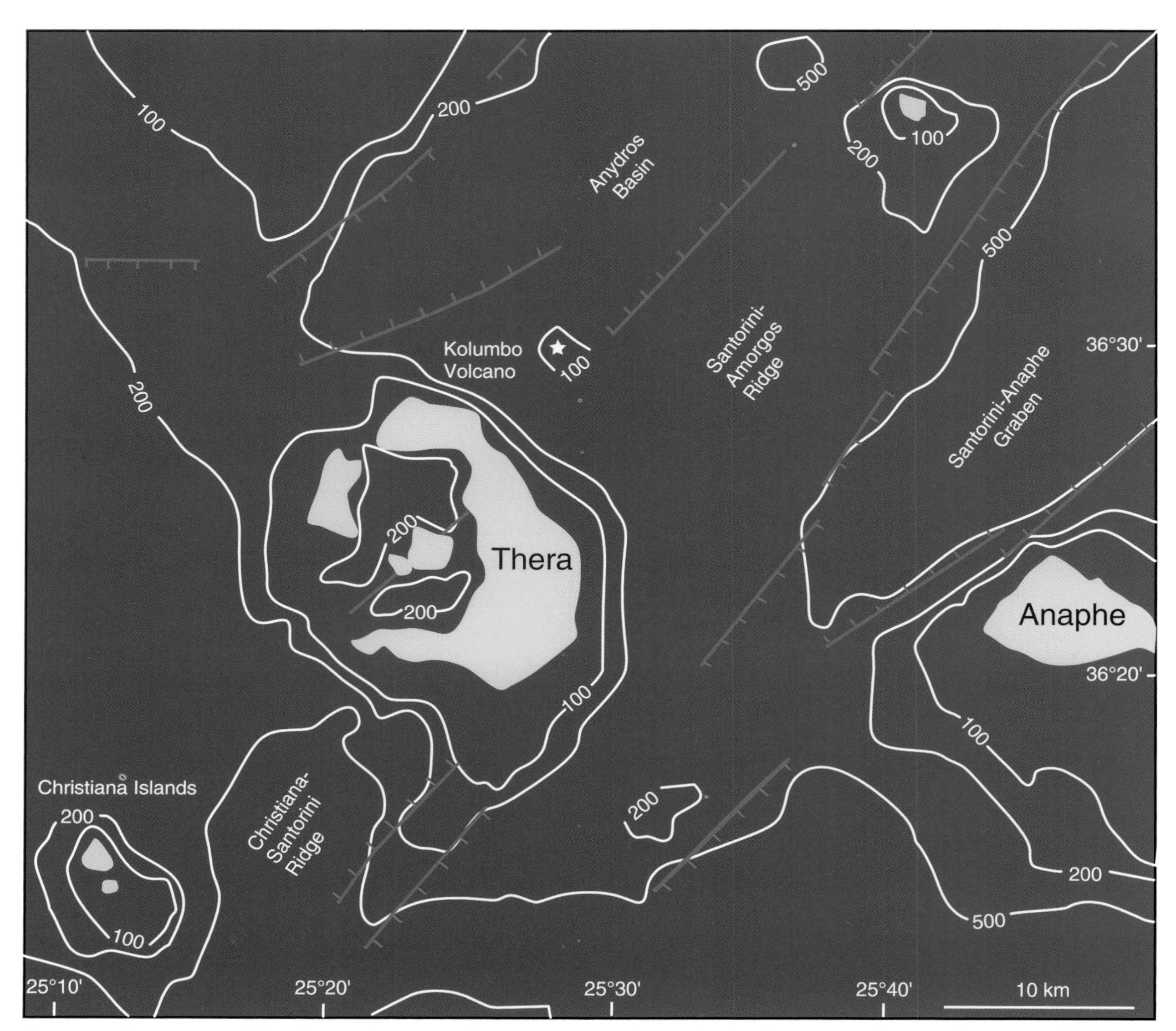

Figure 12.1 The Amorgos Ridge is a northeast-trending tectonic horst (elevated fault block). The islands of Santorini are situated on this submarine ridge together with the Christiana Islands to the southwest and the submarine volcano Kolumbo to the northeast. The Kameni islands in the center of the Santorini caldera are also on the same alignment. Modified after Papazachos and Panagiotopoulus (1993).

enormous pumice eruption that formed the Bu caldera 180 000 years ago. So far as we know, volcanic activity did not begin again until the second century BC, when the island of Hiera was formed. From then on, new lava and ash contributed to refilling the caldera. If this process continues, it could eventually produce another large volcanic complex and gigantic eruption comparable to the earlier ones.

Although the modern Kameni islands were not formed until long after the Minoan eruption, there seems to have been a similar island inside the caldera before that eruption. We refer to it as the 'Pre-Kameni island'. As we noted in Chapter 10, this earlier island was situated in the middle of the flooded caldera, slightly to the north of the present Kameni islands.

The eruption of 197 BC

At the time when the Olympic games were held in Greece for the 145th time, a volcanic eruption occurred in the vicinity of the island of Thera. This was at the time when the Romans concluded an armistice with Phillip III of Macedonia. Strabo said that this was in the fourth year of the Olympic games, which at that time extended over several years. According to our time scale, that would put it in the year 197 BC. According to Livius (*XXXIII.1–13*) peace was made after the battle of Kynoskephalae. This also would be 197 BC.

The name of the new island had a sacred meaning. Several sources confirm that it was called Hiera, which means 'The Holy'. Its appearance had been predicted well in advance by an oracle from Delphi (Box 12.1).

We find accounts of the formation of this island in the works of Plutarch, Pausanias, Justinus and in the Armenian version of the text of Eusebius. The events of that time recorded by Pliny the Elder differ in some respects from the reports of other authors (Choiseul-Gouffier, 1782; von Hoff, 1824; Ross, 1840; Reiss and Stübel, 1868). Pliny the Elder tells us that the island of Hiera also had the name Automate (the self-moving). According to some authors, this island, which later disappeared, formed the so-called Banco platform that was later incorporated into Nea Kameni.

A precursor of Palea Kameni?

Pliny the Elder reports that in addition to Hiera, the island Thia (Theia) appeared during his time, but his dating is almost certainly wrong, because he says elsewhere that an island appeared in the neighborhood of Hiera in the year 19 AD. This report is probably based on a transcription error since others do not mention it. Moreover, he writes about an island with the same name in the Eolian Islands of southern Italy.

The eruption of 46 AD

When the city of Rome was celebrating the 800 years that had passed since it was founded, the island Thia (Theia) – the Godly – appeared. Its appearance at this time is reported by several authors including Seneca, Dio Cassius, Aurelius Victor, Orosius, Cassiodorus (Box 12.1), so the eruption of 46 AD is very well dated.

Seneca says, 'in our time a new island appeared in the Aegean Sea between Thera and Therasia'. Cassiodorus (468–562 AD) wrote; 'in the fifth year of the reign of the emperor Claudius when Vinicius and Cornelius were consuls there appeared between Thera and Therasia an island of 30 stadia'. According to our time scale that would be in 46 AD. In speaking of 30 stadia, he must have been referring to the island's circumference, for 30 stadia equals 5.6 kilometers. Victor Aurelius, who lived in the fourth century AD, adds further details about this eruption. In his *Historia Romana* he says that in the sixth year of the reign of the emperor Claudius and in the year of the 800th anniversary of Rome (AD 47) an island appeared in the Aegean Sea at the same time as an eclipse of the moon. The latter observation enables us to get the date of the appearance of the island even more accurately. As already reported by Labbeus (1670) this eclipse of the moon occurred on 31 December 46 AD between 1700 and 2100 hours. This eclipse is also reported in the *The Canon of Darknesses* of Oppolzer (1887).

It is possible that the report by Philostratos, who lived in the first century after Christ, can be interpreted in terms of the same event. He reports an earthquake on Crete and the birth of an island at the same time between Thera and Crete (see Box 13.1).

The island of Thia could be a part of a reef that later was called Banco.

A historical turning point (726 AD)

In the time span of 46 to 726 AD there are no reports of volcanic eruptions in the caldera. This could be for two reasons. Either there were no eruptions or no reports were preserved. However, Fouqué (1879) points out, quite correctly, that an eruption as strong as that of 726 normally comes only after a long period of repose.

Box 12.1 Historical sources dealing with the origin of the Kameni Islands partly after Richard (1657), Goree (1707), von Hoff (1824), Ross (1840), Reiss and Stübel (1868) and Dobe (1936).

The geographer Strabo (66 BC–24 AD) described the formation of what was certainly the island of Hiera in 197 BC. His text and that of Seneca are quite similar, since, as Ross (1840) noted, both used the Greek philosopher Posidonius (135 BC–51 BC) as their source.

Midway between Thera and Therasia fires broke forth from the sea and continued for four days, so that the whole sea boiled and blazed, and the fires brought up an island which was gradually elevated as though by levers and consisted of a burning island twelve stadia [2.2 kilometres] in circumference. When the eruption was over, the Rhodians, [about 167–200 BC, as we know from Livius (31.15) and other sources] at the time of their maritime supremacy, were first to venture upon the scene and to erect on the island a temple in honor of Poseidon Asphalios.

Strabo (*Geography*, I. 3. 16)

In the same year [197 BC] an earthquake occurred between the Islands of Thera and Therasia while before the very eyes of the astonished sailors an island and hot water suddenly arose from the depths of the sea.

Justin (Justinus, Marcus Justinianus, 2. Century AD) (*Trogi Pompei Historiarum Philippicarum epitoma*, 30.5)

Does anyone doubt that air brought Thera and Therasia into the light of day, as well as that island which in our time was born before our very eyes in the Aegean Sea?

Seneca (*Natural Questions* VI, 21. 1)

According to Posidonius, an island arose in the Aegean Sea, in the tradition of our forefathers. The sea foamed during the day and smoke was carried up from the depths. Finally night brought forth fire, not a continuous fire but one that flashed at intervals like lightning, as often as the heat below it overcame the weight of water lying above. Then rocks and boulders were hurled up. The air had expelled some of them before they were burnt, and so they were undamaged while others were corroded and changed to light pumice. Finally, the top of a burned mountain emerged. Afterwards, the rock gained height and grew to the size of an island. The same thing happened again in our own time during the second consulship of Valerius Asiaticus.

Seneca (*Natural Questions* II, 26.4–6)

Valerius Asiaticus was consul in the year 46 AD.

Pliny the Elder: Latin in full Gaius Plinius Secundus (AD 23–79)

The following text by Pliny is obviously not complete since it does not make sense either historically or geographically:

The famous islands of Delos and Rhodes are recorded in history as having been born from the sea long ago, and subsequently smaller ones, Anaphe beyond Melos, Neae between Lemnos and the Dardanelles, Halone between Lebedos and Teos, Thera and Therasia among the Cyclades in the 4th year of the 145th Olympiad; also in the same group Hiera, which is the same as Automate, 130 years later; and 2 stades from Hiera, Thia 110 years later, in our age, on July 8 in the year of the consulship of Marcus Junius Silanus and Lucius-Balbus.

Pliny (*Natural History*, II. LXXXIX, 202)

Plutarch

Plutarch (AD 46–119), in listing events that were forecast and subsequently took place, specified the date and place of the eruption. As an example of a successful prediction, he cites the rise of an island between Thera and Therasia.

But when the offspring of Trojans shall come to be in ascendance

Over Phoenicians in conflict, events shall then be beyond credence;

Ocean shall blaze with an infinite fire, and with rattling of thunder

Scorching blasts through the turbulent waters shall be driven upward;

With them will come a rock, and the rock shall remain fixed in the ocean,

Making an island unnamed by mortals; and men who are weaker

Shall by the might of their arms be able to vanquish the stronger.

The statement of the oracle at Delphi was interpreted in this way:

Then, within a short lapse of time, the Romans defeated Hannibal and consequently later Carthage, Philip was defeated by the Etolians and Romans, and finally, an island rose from the depths of the sea in the midst of violent boiling of the waves and ejections of burned debris. Who would dare to say that all this is coincidence? Does this not demonstrate the truth of the prophecies?

Plutarch (*De Pythiae oraculis*, vol. VII, p. 267, edition Loeb 1936)

According to Titus-Livy (XXXIII.1–3), the treaty of peace between Philip and the Romans was signed after the battle of Cynocephale in 197 BC. Therefore, this same date must be assigned to the appearance of Hiera. It is true that the same author, in Book XXXIX, Chapters 46 and 47 of his history, speaks of a treaty concluded between Philip and the Romans and says that this took place under the consulate of Claudius Marcellus and Quintus Fabius Labeo, who administered the republic during the 570th year after the founding of Rome (184 BC). But the treaty in question here is evidently later than the one that followed the battle of Cynoscephalae and was concluded under the consulate of Quintus Flaminus. The date of the latter, 197 BC, is closer than that of the second (184 BC) to the year 201 in which the second Punic War took place, and since Plutarch considered the conclusion of the treaty with Philip and the end of the second Punic War as having taken place within a very short time, it follows that the first of these two dates must be favored over the second as corresponding to the formation of Hiera at the same time as the two events cited above.

The eruption of 726 was important because it produced a new island on the north side of Hiera that was later united to form the foundation of the present Palea Kameni. More important, however, were the historical consequences of this eruption (Box 12.2). The eruption was so strong that pumice covered the sea as far as the town of Abydos in the Straits of Dardanelles. Ash was also reported on the coast of Asia Minor and Macedonia, as is reported by Nicephoros (758–828) and Theophanes (752–818).

Despite the importance of this eruption we find only few traces of the pumice on Santorini itself. Ross (1840) speculated that the conspicuous young lava tongue at the north end of Palea Kameni dates from the eruption of 726 (Fig. 13.0), and von Seebach (1868), Fouqué (1879), and Fytikas *et al.* (1990a) shared this opinion.

The break up of Palea Kameni (1457–1458)

A map of Santorini by Buondelmonte (1465–1466) has a Latin inscription noting that eight years earlier an island had appeared between the islands of 'Santilini' and 'Thirasia'. This would place it in 1457 or 1458 (Fig. 12.2).

The same conclusion can be drawn from an inscription carved beside the portal of Skaros castle. The original, which was engraved on a marble plate in a church in Skaros castle, is no longer preserved but the Jesuit priest François Richard (also called Ricardus) reproduced the text in his book in 1657. Also, Athanasius Kircher has in his *Mundus Subterraneus* in 1665 a short account by Richard about eruptions on Santorini, which also contains the Latin text as follows:

Magnanime Francisce, Heroum certissima proles,
Crispe vides oculis clades, qui mira dedere,
Quinquies undenos istis jungendo duobus,
Septimo Kalendas Decembris, murmure vasto

Box 12.2 The historical consequences of the 726 AD eruption.

The oldest reports about the eruption in the year 726 AD are given by Nicephoros (758–823), who was at that time patriarch of Byzantine. But the historian Theophanes (752–818) also gives a report about the same in his *Chronographica*. Later, Cedremus (after 1059) repeated the report in almost the same words. It is also known that the Emperor Leo III started the 'Iconoclast Controversy' over the Iconoclasts, who at that time were destroying religious images (icons) and monasteries.

According to Nicephoros there is a connection between the Iconoclasm of the time of Leo III and the eruption in the year 726. Leo III took the eruption as a sign of God's wreath over the admiration of images. He removed, for example, an icon of Christ in the town of Chalke. In retaliation, Pope Gregory II excommunicated the Iconoclasts, which led to a breakdown of relationships between Rome and Byzantine that changed the course of history in the Western World. When the Arabs invaded Europe, the Pope, who wanted to defend Christianity and to stop the Muslim invasion, could not call on the Byzantine church for help. Under the leadership of Charles Martel, who was at that time the Carolingian mayor and *de facto* ruler of the Frankish kingdom, he was able to stop the invasion of the Muslims at Poitiers in France in the year 732 – six years after the eruption of Palea Kameni.
Nicephoros says on page 64:

We should not forget what happened at that time (AD 726) in the Cretan Sea near Thera and Therasia. In the beginning of the summer dark smoke was seen rising from the depths of the sea along with flashes of fire. Then masses of stone were observed to join with the island called Hiera, which also had arisen from the depth, as had also the islands Thera and Therasia. The enormous masses of stone covered the entire surface of the sea as far as Abydos and even further on the coast of Asia Minor. Wherever these masses spread the water was so warm that no one could touch it. When the Emperor [Leo III] became acquainted with this happened; he guessed it was a sign of the wrath of God and began to inquire into what could have caused it. He came to the conclusion that it was necessary to repeal the holy image-worship as he thought, by mistake, that the prodigy had been caused by the public exhibition and veneration of such images.

(*Scriptorum Historiae Byzantinae*, Bonnae 1837)

Theophanes wrote in a more precise and detailed way:

In the same year (726) during the summer season; a vapor was seen to boil, just like a fire from a chimney, for some days in the depth of the sea, between Thera and Therasia isles. As the fire had increased and extended a little at time, a smoke colored as the fire appeared everywhere. Then a large quantity of pumice stones with a consistence of terrestrial substances scattered in form of big piles all over the Asia Minor, Lesbos, the town of Abydos and the Macedonian coastline region so that the whole sea surface had been covered with pumice floating on the water. In the middle of a so vast fire an isle (was formed) made up of terrestrial substances which did not exist before and which joined again the isle that is called the Holy one. In the same way, like the already mentioned isles, that is to say Thera and Therasia, one time boiled so (were formed) now this (new) isle in the time of the Emperor Leo. He feared that the wrath of God was shown against him and he stirred up, with arrogance, the struggle against the holy images, which were worth of worship.

(1655) (Translated from the original Greek text by M. Fytikas)

Vastus Theresinus immanis saxa Camenae
Cum gemitu avulsit, scopulusque a fluctibus imis.
Apparet, magnum gignet memorabile monstrum.

Dr. Mary Jaeger translated the Latin as follows:

Great-hearted Franciscus Crispus, surely an offspring of heroes,
you see the destruction that has shown us wonders,
with fourteen hundred and fifty-five of Christ's years
slipping by, and adding two to that number,
Seven days before the Kalends of December, with a great rumble,
immense Theresinus (the sea around Therasia) tore enormous rocks from Kameni,

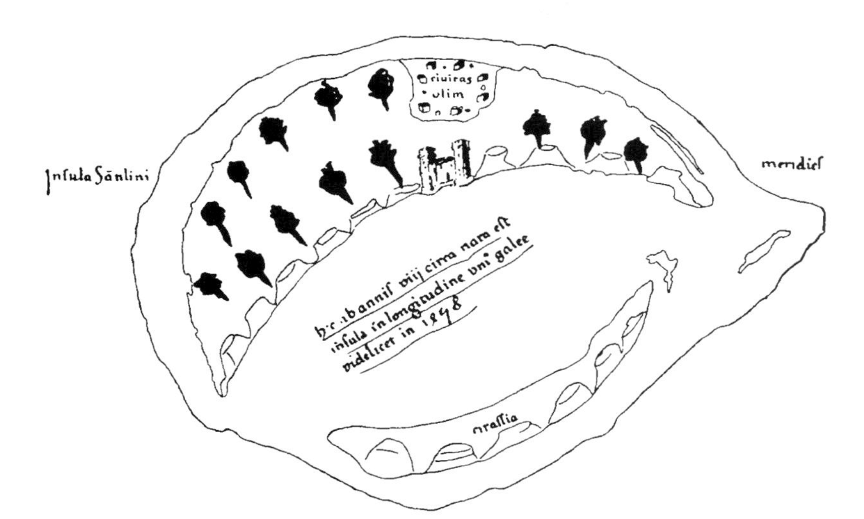

Figure 12.2 A map of Santorini by the mapmaker and traveler Buondelmonte, from the year 1465 to 1466, contains a very schematic drawing of the islands of Thera and Therasia. One sees among other features the castle of Skaros, which no longer exists. The Latin text written in the middle of the caldera says that in the year 1458 an island appeared that had a length of one 'galee'. A galley was a large seagoing vessel up to 160 meters long propeled primarily by oars, and used for war and trading, especially in the Mediterranean Sea, from the Middle Ages to the nineteenth century. From Hiller von Gaertringen (1899).

and with a groan, a crag appeared from the depths of the waves
bringing with it a great and memorable portent.

This text is unclear and hard to translate but seems to state that on the 25th of November 1457 near the Kameni Island (Palea Kameni) either an island disappeared or appeared during the time that Franciscus Crispus was Duke of Santorini. The text includes the first recorded use of the name 'Kameni'. The present island of Palea Kameni has a conspicuous fault scarp on its northeastern side. If the shape and size of the island are compared with those of Nea Kameni, one gets the impression that Palea Kameni is the remaining half of an island, the other part of which disappeared into the sea (see Fig. 1.5).

Another possibility would be that Palea Kameni appeared in 1457 rather than 726 AD and was enlarged by the younger blocky lava now seen on the flatter part of the island near the church of Agios Nikolaos. This is also where the 'Red Water Bay' gets is unusual color from iron-rich warm springs (see Fig. 13.0).

On the map of Pichler and Kussmaul (1980b) the assignment of this blocky lava to the eruption of 1866 (Aphroessa lava) must be an error, because no contemporaneous observer says anything about it. Moreover, the lava tongue is shown on maps that were made before the eruption of 1866.

Mikra Kameni appears (1570–1573)

From 1570 until about 1573 a new island was formed in the sea about four kilometers northeast of Palea Kameni. Having risen with 'fire in the sea', it was given the name Mikra Kameni – small burned island. It had a crater from which coarse ash and lapilli were ejected. Athanasius Kircher, who refers to the account of the Jesuit priest Ricardus, reported the event. The botanist Tournefort, who visited Santorini in 1700, also reported the appearance of Mikra Kameni from the account of Father Richard (Fig. 12.3). The eruption of 1570 is also mentioned in a Greek manuscript that also describes the eruption of the year 1650. The archaeologist Ludwig Ross (1840) published a summary of this manuscript.

The activity shifts

The Kolumbo eruption of 1649–1650

On 27 September 1650, following a year of frequent earthquakes, smoke and steam rose from the sea about seven kilometers northeast of Cape Kolumbo on the east coast of Thera. A pumice eruption soon

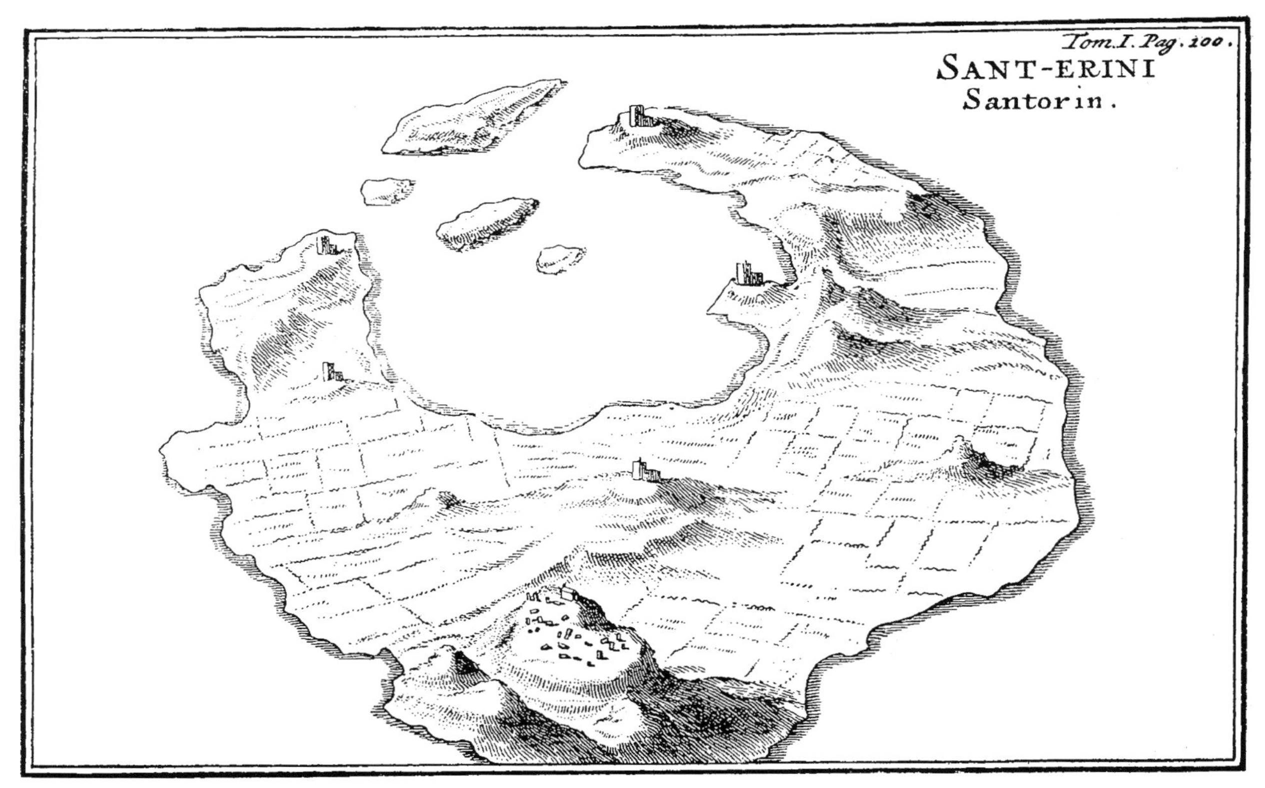

Figure 12.3 The first reasonably accurate overview of Santorini shows the caldera with the islands of Micra and Palea Kameni. On the main island one sees the ruins of Ancient Thera as well as the five fortresses on elevated parts of the island. These are (clockwise and starting from top): Apano Meria (today Oia), Skaros, Pirgos, Emborion and Akrotiri. The drawing was made by the botanist Pitton de Tournefort, who visited Santorini in 1700 (Tournefort, 1714).

followed, sending ash as far as Asia Minor. Four months later, the island that had formed disappeared as suddenly as it had appeared, so that today, the vent of Kolumbo lies 18 meters below sea level (see Chapter 13).

Recent surveys show that the submarine cone is about three kilometers long and has an oval caldera. The deepest point of this caldera is 512 meters below sea level. When the weather is calm, one can see a warm spring near the shore in the neighborhood of Cape Kolumbo.

The effects of the eruption are described in detail by the Jesuit priests François Richard (1657) and Father Goree (1712).

The eruption of 1650 caused more than twenty-five casualties on Santorini. People died of suffocation and toxic volcanic gases (sulfuric acid and probably carbon monoxide). Many animals died and ash covering the fields destroyed most of the year's harvest. A gigantic tsunami hit the eastern coast of Thera. This large wave, which seems to have been triggered by the sudden collapse of the Kolumbo caldera, washed two churches into the sea at Kamari and exposed the ruins of several buildings and the foundation of an older church at Perissa. Marble figurines and other artifacts found in the newly exposed ruins and tombs probably date from the Hellenistic period and early Christian churches.

Ross (1840) left an account of the tsunamis produced by the eruption at Cape Kolumbo and showed that this was a major event. On the island Ios (then called Nios), about 20 kilometers from the site of the eruption, the tsunami reached a height of twenty meters, where it left a layer of pumice carried in by the water.

Along the low, eastern shores of Thera the sea entered far inland, while at Akrotiri and Mavro Rachidi unusually huge lumps of pumice were thrown onto the beach. Karl von Fritsch (1871) mentioned marine pebbles from the east coast of Thera that had been carried by the flood. In 1990 I found such a rounded pebble with marine animals still attached to it at a height of 50 meters in the Potamos valley a short distance east of Akrotiri. The pebble still had calcareous tubes of marine worms on the surfaces. The tsunami carrying the stony material must have been channeled into the funnel-shaped valley.

Ross (1840) collected eyewitnesses accounts of the tsunami of 1650, mainly from Father M. Pègues and the archives of the Catholic church of Fira. This is the most comprehensive collection of original sources that exists about the early stages of formation of the Kameni islands. These include two Cretan poems, the report of a monk from Patmos, and letters exchanged between catholic priests on Santorini and the nearby island of Naxos. From these reports, it is obvious that the flames of the eruption were visible as far as Heraklion on Crete (120 km away) and that pumice reached the island of Leros in the eastern Aegean, where it was seen floating on the sea. The Turks were occupying the city of Heraklion, the largest town on Crete, at the time. It is also reported that ships of the Turkish fleet that had been pulled up on the shore of the small island of Dia north of Crete were washed away by the tsunami. Some contemporaneous handwritten reports of this eruption have been published by Doumas (1978.)

Activity returns to the caldera

The appearance of Nea Kameni

The eruptions of 1707 to 1711 inside the caldera were of great geological interest, because this was one of the rare cases in which a volcano was observed to emerge from the sea. The Jesuit Father Goree gave a detailed description in 1712 (see Boxes 12.3 and 12.4).

On the 21st of May, three days after a strong earthquake, a white island was seen rising from the sea somewhat west of Mikra Kameni. It continued to grow and after a few days people went to the island and discovered that it consisted of pumice and black lava. Marine animals that were still living were found on it. The island grew slowly to a width of 500 to 600 meters and a height of 70 to 80 meters. Later, the sea became very turbulent, hot, and discolored. On the 5th of June, fire was seen and a black island appeared north of the island of white pumice. By the 12th of September the black island was so large that it united with the white one. Explosions and ash eruptions continued until all activity finally ended in September 1711. During the eruption the shorelines on the islands of Thera and Mikra Kameni sank by as much as a meter or more. The new island, Nea Kameni, had

Box 12.3 Observation about the eruptions in the caldera by Father Goree (1712):

During the eruptions of 1707–1711 the castle on Skaros was still inhabited. From there one could observe the growth of the new island. As we learn from Father Goree this went on steadily:

At this time the White island which (as I have said before) seemed to be higher than the Lesser Kameneni [Mikra Kameni] and could then be seen from the first floor of the houses in the castle of Skaros, had now sunk so low, that it could not be far from the second floor.

Father Goree also described enormous flames that might be caused by combustion of erupting gasses – a phenomenon also observed during the eruption of 1866–1870.

The following night there was heard a dull hollow noise, much like that of several cannons shot off at a distance, and at the same time there was seen to rise out of the midst of the funnel flames of fire, which darted very high into the air, and disappeared immediately.

Box 12.4 A new island, Nea Kameni, appears in 1707

The Jesuit Priest Father Goree, eyewitnesses the eruptions of 1707–11 from Skaros Castle. On page 356 he writes:

And it is between this little Island the the Great Kammeni, that on the 23rd of May (New Stile) in the Year 1707, at break of Day, the New Island, of which I am now going to speak, was first discovered.

He continues on page 357–358:

Howsoever it was, some Seamen discovered this Island early in the Morning; but not being able to distinguish what it was, they imaged it to be some Vessel that had suffered Shipwrack, and was driven thither by the Sea. In hopes of making an Advantage to themselves by it, they went immediately to it; but as soon as they found that it was a New Island, they grew afraid, and returning as hastily back again, spread the report over the whole Island; which was the more readily credited, because all the Inhabitants knew, and several of them had themselves seen what happened in the Year 1650.

How great soever the Fright of the Inhabitants of Santorini was, at the first sight of the New Island, yet a few Days after, not seeing any appearance either of Fire or Smoak, some of them more bold that the rest, took a resolution to go and view the Situation of it: Which they did accordingly; and not imagining any Danger, went on Shore upon it. As they had no other design, but to satisfy their Curiosity, they passed from one Rock to another, upon which they met with several very remarkable Curiosities; among which we may reckon a sort of White Stone, which cuts like Bread, and resembles it so well in form, colour, and consistence, that were it not for its taste any one would take it for real Bread. But what pleased them more was a great number of fresh Oysters, which they found sticking to the Rocks; which being very scarce in that Country, by reason of the depth of the Sea, they got as many of them as they could.

While they were busy about this, they perceived the Island move and shake under their Feet. This was sufficient to make them leave it immediately, and to return back faster than they came.

a triangular form measuring 910 meters length in the south, 1650 in the west, and 1440 meters in the east. It reached a height 106 meters.

The Russian monk Barskij, who visited Santorini in 1745, gives us an idea how Santorini looked like after the eruptions of 1707–1711 (Fig. 12.4).

The eruptions of 1707 to 1711 left Nea Kameni with two new bays (Fig. 12.5) which formed excellent natural harbors for the inhabitants of Santorini. The bay in the southwestern part was named Georgios and the other Volcano Bay. In the latter, gases rising through the water gave it a greenish-yellow color. Numerous ships that anchored there found their copper sheeting cleaned of barnacles by the sulfurous water. If such a place were available today, ships that must have their hulls cleaned periodically could save millions.

Later, the hot sulfurous springs were used for health spas. About fifty summerhouses were built on this island, and a small harbor was constructed in the channel of Mikra Kameni. In 1841, the naturalist Edward Forbes visited a small remnant of the white island and collected numerous fossils of marine organisms that normally live in water 30 to 40 meters deep. Forbes gives a list of the findings in his *Report of the Mollusca and Radiata of the Aegean Sea* (1844).

Updoming of the ground such as that observed at the white island is thought to result from subterranean movement of magma and is often a signal for a impending eruption. The doming described here was especially conspicuous for the people of Santorini, who saw a new island rising from the sea. Justin (see Box 12.1) described a similar event.

Doming is rarely as fast and high as it was at White Island, and it normally lasts much longer. When Ross (1840) visited Santorini in 1836 and 1839 there had been no volcanic activity for more than 100 years. He remarked that 'if a new event does not submerge this new island, the volcano will not become extinct for a long while. Fishermen have observed that not far from Megali [Nea] Kameni a sharp reef has started to rise from the sea and grows more and more every year'. About thirty years later a new eruption broke out.

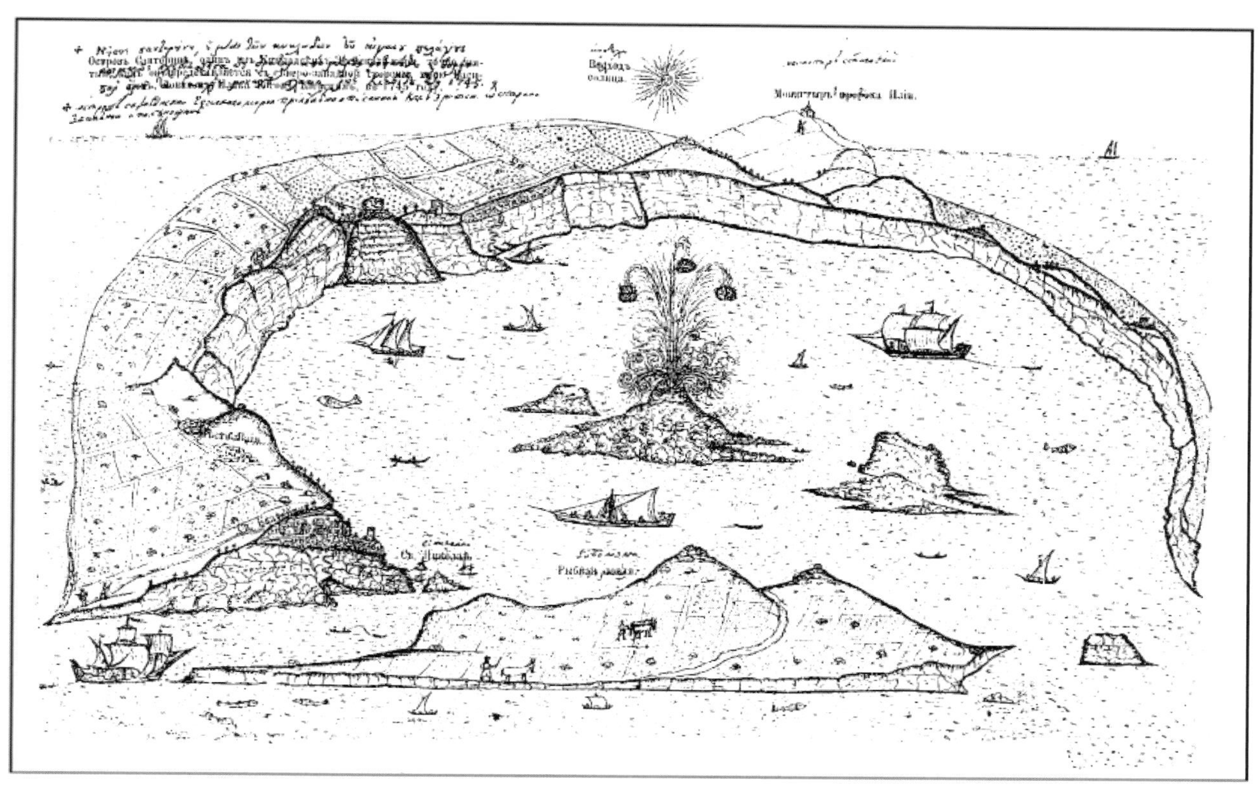

Figure 12.4 The Russian monk Barskij circumnavigated Santorini in 1745. He made astonishingly precise drawings of everything he saw, including geological structures such as the lavas of Skaros. At the time of Barskij's visit there were three Kameni islands in the caldera: Palea, Mikra and Nea Kameni. His illustration shows an active volcano on Nea Kameni. The volcano Georgios was quiet at that time and had had no significant activity since 1711. From Monioudi-Gavala (1977).

Mikra and Nea Kameni unite (1866–1870)

There are far more written accounts of the eruption of Nea Kameni from 1866 to 1870. Among others, those of the naturalists Reiss, Stübel, Schmidt, and Fouqué are most complete. Fouqué provides photos, drawings, and very detailed accounts of these events (Fig. 12.6).

The caretaker who lived with his family on Nea Kameni during the winter, observed on 26 January 1866 that loose blocks were tumbling down the slope at the small harbor between Mikra and Nea Kameni. Cracks were seen in the walls of the houses. In the following days, rumbling and other signs of impending volcanic activity continued to accumulate: the sea began to boil in Volcano Bay and sulfur-rich gases rose through the water. At the beginning of February, fumes were rising and dark blocks of lava began to move slowly on the surface of the water near Volcano Bay. Lava that was already cool appeared in the cone and slowly moved away from the eruptive center. The center was named 'Georgios' after the king of Greece. The products of the eruption reached Nea Kameni on 6th of February and covered the summerhouses near the small harbor (Fig. 12.7). Later, a new vent called Aphroessa opened on the southern edge of Nea Kameni, and its lava joined that of Mikra Kameni on March 1866. On 20 February Georgios had a strong explosion followed by an ash eruption that continued at short intervals for many months. The geologist Karl von Seebach described the grand display he witnessed from an observation point close to the eruption (Fig. 12.8, Box 12.5).

Figure 12.5 A view of the caldera as seen from the fortress on Skaros shows three young volcanic islands. These are Mikra Kameni, Nea Kameni in the middle, and Palea Kameni in the rear. On the right side can be seen the island of Therasia. These islands provided natural harbors that were used until the eruption of 1866 to 1870 (From Choiseul-Gouffier 1782, plate 14).

On the 10th of May, a new vent opened forming some small islands of black lava that were named the 'Maionisi' (Isles of May). They soon disappeared, leaving some shallow areas under the sea that can be seen even today when the sea is calm. The eruption continued until the 5th of October 1870. In the following years only small gas eruptions and fumarolic activity were seen on Nea Kameni, but activity returned in the twentieth century after five decades of quiet.

Eruptions in the twentieth century

There are only few descriptions of volcanic eruptions that are more detailed than those of Nea Kameni in the twentieth century, possibly because it was easy for volcanologists to observe all changes on the island from the security of the high caldera rim on Thera.

The eruptions of 1925 to 1928

Following the eruptions of 1866 to 1870 a period of 55 years passed before another eruption broke out, in August 1925. The Greek government sent the geologists George Georgalas and Nikolaos Liatsikas to Santorini to observe the eruption. In September 1925 a group of volcanologists consisting of the Germans Hans Reck and Friedrich Dobe, and the Dutch Maur Neumann van Padang joined them. Their observations of the eruption are very thorough (Reck, 1936). Additional information can be found in the work of

Figure 12.6 The eruptions of 1866 to 1870 created the Georgios volcano and caused the smaller Mikra Kameni to be joined to the larger Nea Kameni. These two drawings from the book by Fouqué (1879) show the island as seen from Fira.

Figure 12.7 The eruption of 1866 to 1870 on Nea Kameni destroyed about 50 houses and the only good harbor of Santorini. Since this event, no further houses have been built on this active volcanic island. This drawing of the 6th of April, 1866, shows the destroyed small harbor called Volcanos in a view from Mikra Kameni. From von Seebach (1868).

Box 12.5

The eruption of 1866 on Nea Kameni as described by Karl von Seebach (1868):

The boat is already waiting and the instruments have been carried down in advance. We descend the steep path and sail to Nea Kaymene [Kameni]. Just before reaching Mikra we pass a reef, where bigger ships can anchor. The water here is only six fathoms deep. We sail along the southern edge of Mikra-Kaymmene until the area of most recent destruction lies before us. The abandoned and destroyed houses rise sadly out of the mass of dark lava blocks. Behind them Georgios rises to a height of about 150 feet, a sad field of devastation where the blocky slopes contrast sharply the white steam emerging from the fissures everywhere and uniting as it rises into a single column of steam. The water around the boat is getting warmer and is carried in strong current from the source of heat. Driven by the wind, smaller steam clouds are dancing over the sea in miniature cyclones. The thunder of the pulsating activity gets louder and stronger. Finally we land at the destroyed houses on the pier of the small harbor after a sailing trip of about half an hour and walk along the ejected blocks to the foot of Georgios volcano, with the intent to climb up. This is not simple, however. Individual blocks are piled loosely over each other and often a slight touch is enough to upset their balance. They tumble down the steep slope causing others to fall with them. Their edges are sharp and abrasive. Our hands are bloody from many small wounds and even strong boots are cut. From time to time one hears a loud cracking sound as from a rapidly cooling oven, followed by a light clinking like breaking porcelain. It is the lava that is cooling and shrinking beneath us and, in small pieces of glassy material, falling down the newly formed fissures. Finally we reach the top. We are standing on a slightly domed platform on which the glowing air is shimmering so that dancing objects in the background are scarcely recognizable. Here the blocks are even bigger than at the edge, and many are bleached by gases escaping from open fissures. We advance slowly feeling the way to avoid hot, glowing spots that are not visible during the day. At times the erupting steam prevents us from seeing objects at close range. We work slowly through many detours to the place where strong jets of steam are emitted with a great roar. Most of it is obviously water vapor, since one can breathe fairly freely. Only here and there can one detect a slight whiff of sulfuric acid.

The heat is increasing more and more and we finally reach a broad fissure that gives off much heat and burning flames. We are unable to go on. Heat is also given off from the still moving incandescent lava that fills the depths of the fissure. It is still visible in the darkness of the night. For this purpose we climb to the top of Nea Kameni where one can watch the whole eruption. Georgios is situated at the south foot of the cone, which is surrounded on the east and west by two large solfatara fields where yellow sulfur had been sublimated. Rocks at the summit are bleached by great quantities of gas emitted from numerous, large, crosscutting fissures. There is no real crater, however. The summit is clearly visible during the periods of relative calm between pulses of stronger clouds of wildly rising steam. Seen from a distance, Aphroessa looks like a mound of dirt made by a mole. Like Georgios, it has no crater, but everywhere fumes are being emitted from between the larger blocks. The fissures are not as wide as they are on Georgios but they are colored a bright cinnamon brown and from time to time chlorides can be recognized in them. Only seldom does one observe on Aphroessa pulses of activity like those on Georgios when large amounts of steam break through with great power.

At dusk, the scene starts to change. The acid bleaching of the main fissures on Georgios seem to glow in the dark, and on Aphroessa the red incandescence is visible everywhere. With total darkness the glowing dust clouds have a more intense light and have increased in number. The dark column of steam over Aphroessa is now a great flame and with every pulse the rising steam-clouds from Georgios are illuminated. The most spectacular and interesting phenomenon is the burning flames that break out of all the fissures. This rare, much-discussed phenomenon was clearly recognized by all scientists who have studied the eruption of 1866. The flashing flames ascended with growing intensity and accelerated with each pulsation. Its core was bluish white, its margins carmine red. One can exclude the possibility that this was just a reflections, because both the flame and its reflection could be seen clearly side by side, and they were indistinguishable.

Figure 12.8 The eruption of Nea Kameni in the years 1866 to 1870 presented a spectacular display that was seen by numerous spectators. The fireworks were especially grandiose during the night. From *Illustretet Tidende* (1866), Copenhagen.

Figure 12.9 The four-year eruption of Nea Kameni from 1925 through 1928 produced two domes, 'Daphne' and 'Nautilus'. The latter is named for the ship of Captain Nemo who, in Jules Verne's *20,000 Leagues under the sea* emerged in this volcano. Photo Nellys, courtesy Professor A. Kontaratos.

Figure 12.10 The schematic diagram shows the development of the island of Nea Kameni during the past four centuries.

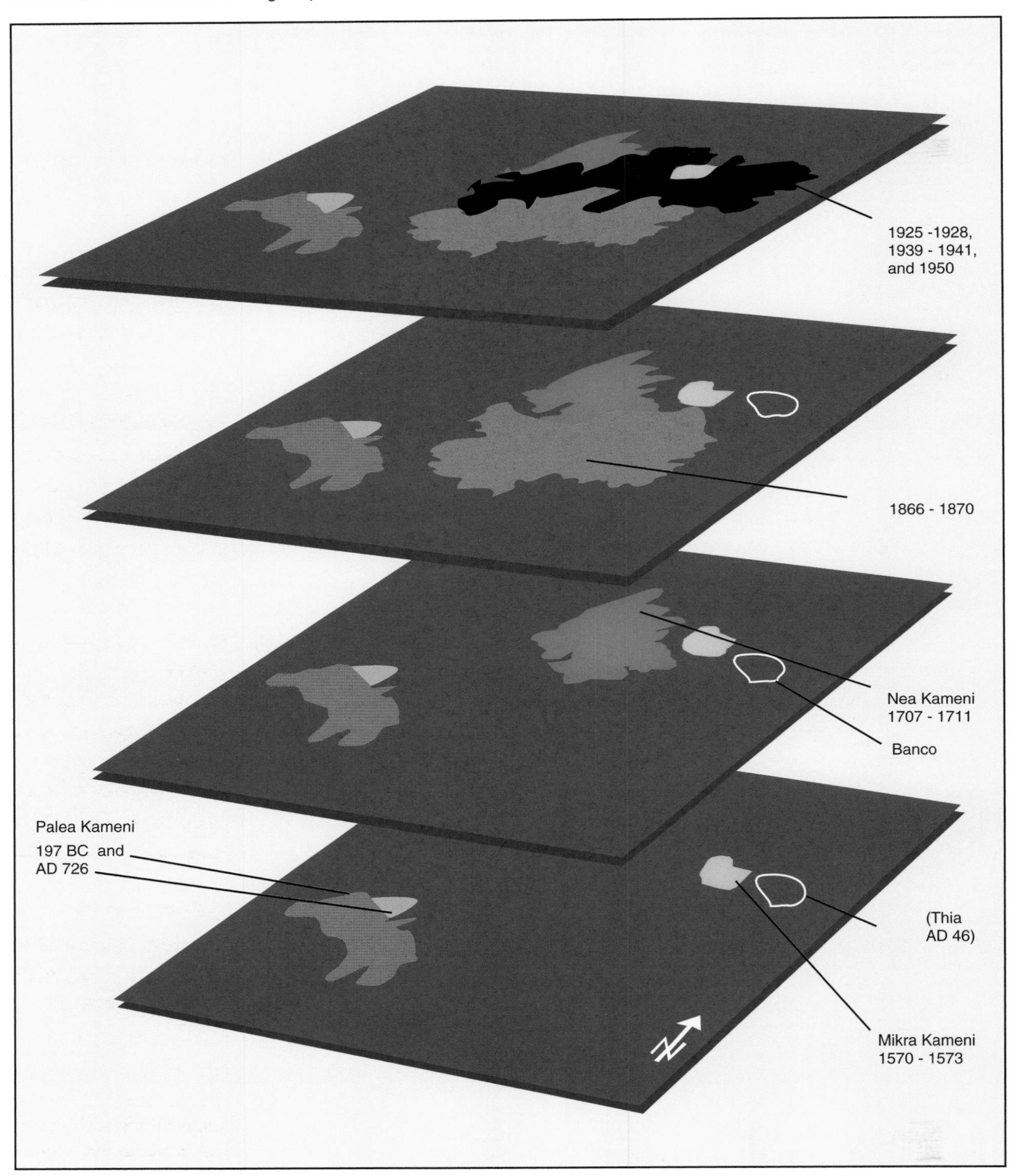

Georgalas and Liatsikas (1926), Ktenas (1926, 1927), Ktenas and Kokkoros (1928), and Washington (1926), all of whom described the course of the eruption. A summary of the literature and observations was given by Georgalas (1962).

The activity started with a marked increase of temperature in the Bay of Red Water (Kokkina Nera) accompanied by sinking of the east coast of Nea Kameni. Soon after this, steam and hot water rose and at the same time the first lava emerged. The activity later shifted and formed the dome 'Daphne' (Fig. 12.9). An explosive phase then began sending a phreatomagmatic eruption column to heights of as much as 3.2 kilometers. This explosive activity continued until January 1926, when all activity stopped for four months. The style of eruption then changed again; pyroclastic flows were triggered by interaction of the magma with water. A second pause lasting from May 1926 to January 1928 was followed by four phreatic eruptions. Finally, an effusive–explosive phase followed, in which the Nautilus (Naftilos) dome was formed. The latter name is taken from the book of Jules Verne, *20,000 Leagues Under the Sea*, in which the submarine *Nautilus* rose in the Bay of Santorini. At the end of the eruption the northern 'Liatsikas lava stream' filled an area of 225 500 square meters and the eastern 'Georgalas lava field' an area of 610 000 square meters.

The eruptions of 1939 to 1941

Upheaval of several areas and subsequent effusions of viscous lava alternating with explosive phases characterized the eruptions on Nea Kameni in 1939 to 1941. The vents, domes, and lava fields were named after volcanologists who had worked on Nea Kameni earlier.

Two of the domes were named after Fouqué and Ktenas, while two smaller ones were named Smith A and Smith B. A fifth (Reck) rose on July 17 1940 and was followed by a lava flow. In November 1940 the 125 meter high Niki dome appeared and produced three lava flows covering an area of 170 000 square meters. The eruptive cycle came to an end in July 1941.

The eruptive activity of 1950

The eruption of 1950 consisted of phreatic explosions and small extrusions of viscous lava. One of the lava flows was named for the Greek geologist Liatsikas (see Georgalas, 1953). The 'Reck dome' disappeared, leaving only a huge funnel-shaped depression. This probably happened when a strong explosion was heard from Nea Kameni. Georgalas (1959) describes this eruption in full detail.

We can be sure that this eruption was not the last word in the volcanic story of Nea Kameni. The island has experienced an enormous increase in volume during the twentieth century and will probably continue to do so in the future (Figs. 12.10 and 12.11).

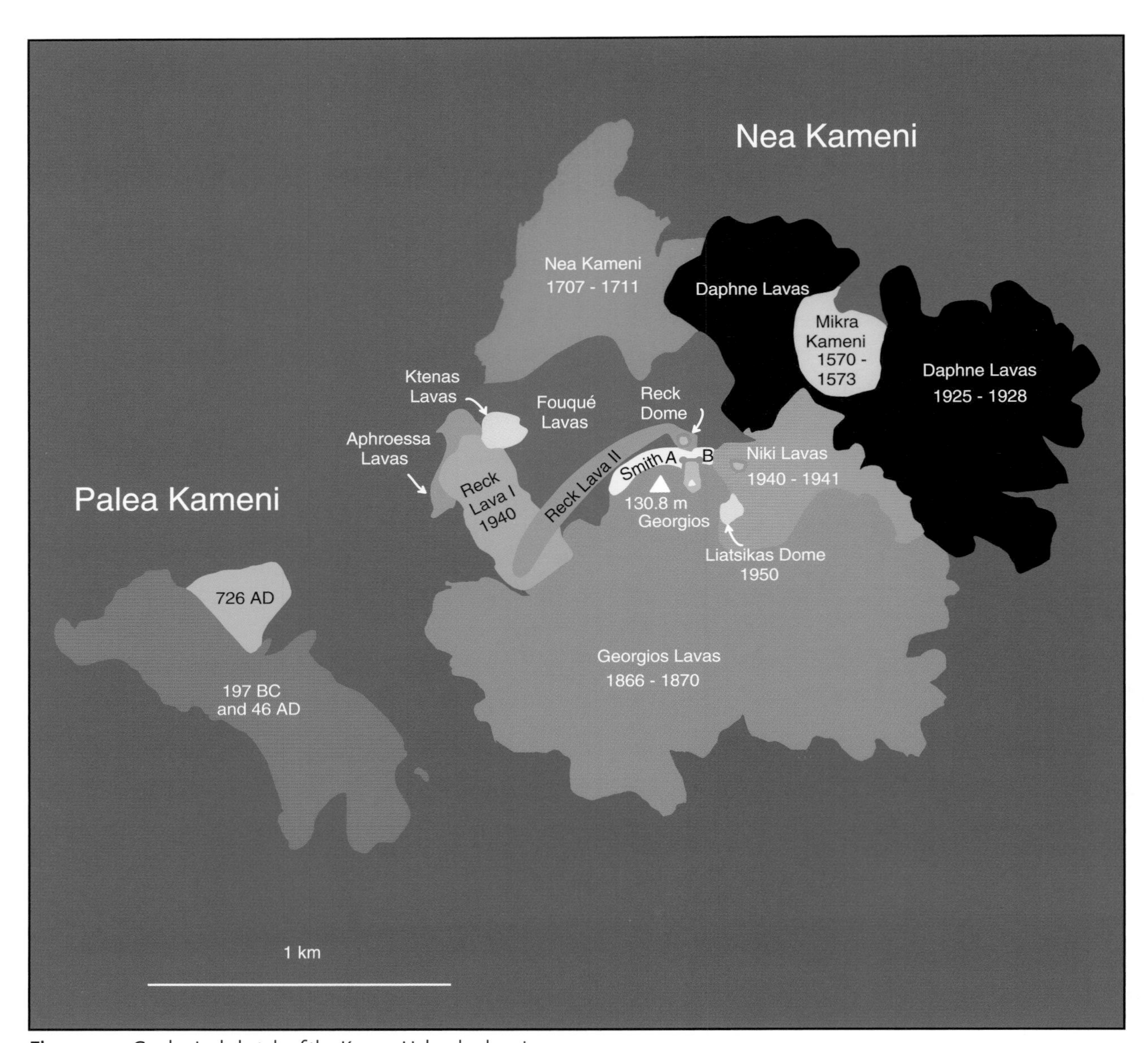

Figure 12.11 Geological sketch of the Kameni islands showing how the eruptive centers on Nea Kameni follow the same northeasterly trend extending from the Christiana islands to the Kolumbo volcano. Modified after Georgalas (1962).

13

Volcanism today

A dormant volcano can return to activity at any time and cause unpredictable damage. What can one do to forecast an volcanic event and how can one reduce its threatening effects?

On Santorini there are numerous signs indicating that volcanic activity is still not finished and can break out in the future just as it has in the past (Fig. 13.1). The primary factor governing its state is heat. Hot springs are found at several places on Thera, and fumaroles are quite active on Nea Kameni. The earth-

Figure 13.0 On Palea Kameni the Red Water Bay near the church Agios Nikolaos is very conspicuous owing to the iron-rich sediments formed in the warm sea water. In the upper part of the picture one sees a dark tongue of blocky lava from an eruption in AD 726.

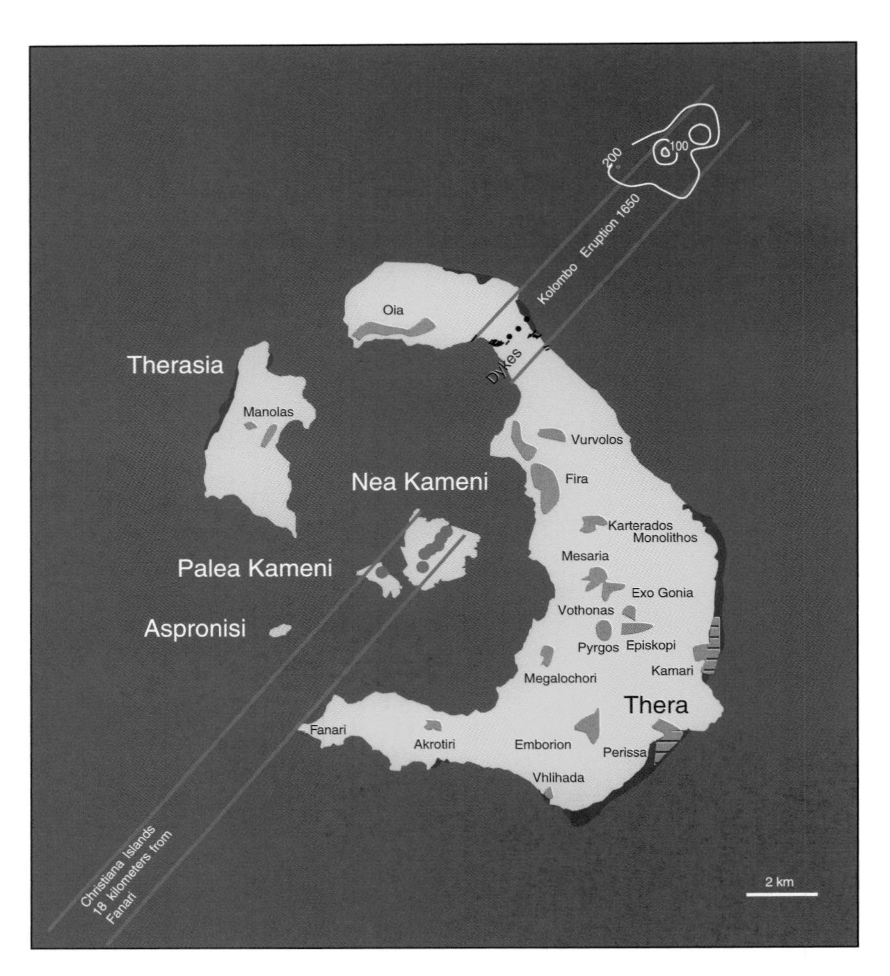

Figure 13.1 On Santorini two dangerous zones (red stripes) are marked by concentrations of volcanic activity. One northwest-trending zone includes the eruptive centers of the last two centuries on the Kameni islands and another, shifted slightly to the north, runs through the volcano Megalo Vouno and the submarine volcano of Kolumbo. Both are part of a larger tectonic zone of weakness stretching from the Christiana islands in the southwest to the Kolumbo volcano in the northeast. Most future eruptions are likely to be on these tectonic lines. The main areas endangered by tsunami are the flat-lying coastal areas of Thera and Therasia (dark blue on the map). These are the same areas affected during the eruption of the Kolumbo volcano in 1650. The figure has been complied from Heiken and McCoy (1984) and Fytikas *et al.* (1990a).

quakes that occur from time to time also remind the people that the ground on which they have built their houses is still active.

Fumaroles, solfataras, and warm springs

Hot steam and yellow and white sublimates on the rim of the Georgios crater show that magma still resides beneath Nea Kameni. At the bottom of the steam vents temperatures are as high as 95° C, and sulfur crystals form a yellow crust on the altered rocks. Fumaroles and solfataras such as these are the visible signs of post-volcanic activity. Thermal springs occur in the sea around Nea and Palea Kameni; when the sea is calm they are marked by the reddish brown color of iron oxides. The sea water in their vicinity is unusually warm. Throughout Santorini, the remains of prior volcanic activity are almost ubiquitous.

For instance, there are warm springs on the coast near Cape Kolumbo and in the sea where the Kolumbo eruption of 1650 occurred. They are also found on the southern cliffs of the Platinamos ridges at Vlihada and in the caldera at Cape Plaka (Fig. 13.2) and Cape Therma (Fig. 13.3). Kircher (1665), Tournefort (1707), and the Jesuit priest Pègues (1842) mentioned the latter two sites, which they considered to be healthy springs. Pègues obtained analyses of the chemical composition of the water at Plaka through the chemist Lauderer. He lists the sicknesses that had recently been cured there and speaks of the many tokens of thanks given to the Virgin Mary by people cured by the waters of Plaka.

Ore minerals in warm springs

A bay on the north side of Palea Kameni is conspicuous because its water has a deep reddish-brown color (Fig. 13.0). In this shallow area, the sea bottom is covered with a layer of mud up to 60 cm thick, the uppermost part of which is reddish brown. The bay is called the 'Red Water Bay'. The water is about 38 °C and nearly everywhere in the bay bubbles can be seen rising through the water.

Today, the bay is a favorite spot for tourists who like to sit in the warm mud, but geologists find it interesting too, because it provides an excellent opportunity to study the formation of metallic ores in a marine

Figure 13.2 The warm springs at Plaka and Thermia have been known for centuries as health spas. Both are related to the unconformity between the metamorphic basement and the overlying volcanic rocks. The church of Panagia can be seen in this view.

Figure 13.3 Near the Christos church, at Cape Thermia, a warm spring is located on the oblique unconformity seen in the lower part of the picture. A steep path leads down from the town of Megalochori to the sea.

environment. In this natural laboratory one can learn much about the development of mineralization in subduction zones. The geochemical studies on Palea and Nea Kameni (Boström *et al.*, 1990) support the hypothesis that hydrothermal systems in young volcanic arc and subduction zones are capable of producing ore deposits. This is in contrast to hydrothermal areas in continental environments and old island arcs.

Why do we have ore minerals in these places? Investigations by Puchelt *et al.* (1973) show that the warm reactive gases rising from great depth dissolve iron from the rocks they rise through and deposit it on the sea floor. There it reacts with the sea water and is precipitated as amorphous (i.e. uncrystallized) iron hydroxides. These places have gas emissions rich in carbon dioxide. Large amounts of iron carbonate, siderite, are also precipitated. The iron formation occurs in association with iron bacteria of the species *Gallionella ferruginosa*.

Until a few years ago it was thought that *Gallionella ferruginosa* is responsible only for bog-iron deposits laid down in fresh water, but more recently, new analyses and experiments have shown that *Gallionella* is also active in marine environments. Optimum conditions for it are at low oxygen concentrations and acidity with relatively high concentrations of hydrogen carbonate.

The color of the sediments in Red Water Bay is not uniform. On digging down about 20 centimeters one sees that the mud changes from red–brown to green–black. The change of color comes about when the green–gray iron (II)-hydroxide ($Fe(OH)_2$) and pyrite (FeS) react with oxygen in the sea water to produce the red iron(III)-hydroxide ($Fe(OH)_3$). Further investigation shows that this mud also contains sponge spicules and reworked volcanic material, such as pumice, tephra and volcanic glass.

The iron formation is not confined to this area. Samples taken from the sea floor at several places in the caldera show that iron is being deposited throughout the entire caldera. There is no iron deposition outside the caldera, however. In places, the layers containing iron reach a thickness of three meters, and measurements have shown that in the last 150 years about 350 thousand tons of iron and 19 000 tons of manganese have been deposited (Petersen and Müller, 1978). In addition, lead, zinc, phosphorus, vanadium, chromium, nickel, cobalt and copper are observed locally in low concentrations.

Volcanic hazards

The post-volcanic activity on Santorini is not confined to the regular emission of gases or warm water; earthquakes are also common. People who have lived on a volcano for several thousand years learn to live with this danger and note the distinct advantages and disadvantages of the place they have chosen to build their homes. In the course of the last 2000 years, the inhabitants of Santorini have experienced all varieties of natural catastrophes and plagues and have left reports about their difficulties for their descendants. Accounts of the catastrophes are preserved in sagas, legends, poems and the lessons of preachers (see Box 13.1). When found buried in excavated archaeological sites they enable us to learn what happened. Destroyed settlements, for example, contain a clear message that these people seem not to have taken warnings of impending events seriously.

The experiences of earlier generations are generally forgotten after one or two generations. This is true not only of Santorini but also of all volcanic regions. A volcano that was believed to be extinct suddenly awakens and causes enormous damage to humans and their property. Warnings are almost always too late or misinterpreted.

Today the threats of danger in volcanic regions are closely monitored, but it is not unusual for a volcano that was thought extinct to return suddenly to life. Volcanoes are unpredictable and, even with the most modern instruments, forecasting eruptions is far from precise. Warnings to the people are usually possible only a few hours before an eruption.

An example from Iceland illustrates this problem. In the Krafla area in the northern part of the island

> **Box 13.1 Tsunamis in Roman times**
>
> The Roman author Philostratus described an earthquake on Crete in a biography of Apollonius of Tyana, who became a mythical hero during the time of the Roman Empire in the first century AD. The earthquake was triggered by the formation of an island between Thera and Crete. According to Dobe (1936), it could be the eruption of AD 46, but it could also have been an eruption on the Christiana Islands.
>
> *When Apollonius, as he often did around noon, was speaking with many of the priests of the temple, an earthquake shook the whole of Crete. A thunderous roar did not come from the clouds but from the earth. The sea receded about seven stadia. Most of the people were afraid that when the sea came back it could overwhelm the temple and destroy it. Apollonius told them not to be afraid: the sea gave birth to a new land. Some of the people thought he was speaking of the elements and was saying that the elements could not make changes on earth. After a few days people came from Kydoniatis and reported that on that very day at noon the signal of god was received and a new island was formed in the sea between Thera and Crete.*

monitoring instruments have been installed in order to warn of an impending eruption. It seems, however, that one must reckon with a time span of about four hours from the first sign of movement of magma until an eruption begins. If we were to transfer this condition to Santorini, this time span would be much too short to warn a population of several thousand people and evacuate them from the island.

The volcanic system of Santorini is one of the most active of the Mediterranean region. Thirteen eruptions have occurred there in the last two thousand years. Some of these had a strong impact on the population. Today, Santorini is visited by thousands of tourists, especially in the summer. The number of beds for visitors is now estimated to be 80 000. A strong eruption during the summer would be catastrophic. For this reason the volcano must be monitored by all possible scientific means. A system that uses modern seismological and geochemical monitoring equipment has been installed. Santorini is now a European Laboratory Volcano.

Earthquakes

Earthquakes have probably caused greater damage on Santorini than volcanic eruptions. Seismic records show that earthquakes are of two types, one shallow and the other deep. From time to time the convergence of the Eurasian and African plates triggers earthquakes at depths of 150 to 170 kilometers. These are observed throughout the southern Aegean Volcanic arc. Such earthquakes can have a very strong impact over a large region. In contrast to these tectonic earthquakes, volcanic earthquakes originate at shallower depths and are therefore more local. They are usually caused by movement in the magma chamber. Several old buildings on Santorini show damage that seems to have resulted from earthquakes of this type. They also appear to have affected Cycladic buildings in the excavation at Akrotiri (Doumas, 1983) and of ancient Thera on Mesa Vouno. A Hellenistic temple, which is now the Christian church of Agios Nikolaos Marmaritis at Emborion, is marked by earthquake damage (Fig. 13.4).

The promontory at Skaros had been shaken repeatedly by strong earthquakes. This rock was inhabited during medieval times, because the fortress constructed there offered protection from pirates. From historical sources we know that earthquakes during the eruptions of 1650, 1707 to 1711, and 1866 to 1870 hit the place hard. The Venetian noble families that lived there decided to leave after the strong earthquakes that accompanied the eruptions of 1707 to 1711 (see Box 13.2). According to the sketches and reports (Fig. 13.5) of Count Choiseul-Gouffier, who visited Santorini in 1770, there were still houses on Skaros at that time, as can also be seen in the drawing by Thomas Hope (Fig. 13.6). But when the archaeologist Ludwig Ross visited Santorini in 1836, Skaros had been empty for twenty years. The rich Roman

Figure 13.4 (above and right) One of the many old buildings that show signs of earthquake damage is the former Hellenistic temple that today is used as a Christian church (Agios Nikolaos Marmaritis).

Figure 13.5 (below) This part of a drawing by Choiseul-Gouffier (1782) shows the caldera wall between Skaros (left) and Merovigli. At that time the Skaros hill was still populated. One can also see the bishop's church that not longer exists.

Figure 13.6 In medieval times the Skaros promontory was a fortress with a densely populated, catholic settlement. This drawing by Fauvel from the work of Thomas Hope (1769–1831) shows the buildings crowded around the fortress. A note by E. M. Leycester (1850, pp. 22–23):

(November 1st): I am now seated in front of the ducal castle of Skaro; my position is under a huge step of red lava and slag, which supports a part of Merovouli [Merovigli]. The promontory of Skaro is separated from me by a high ridge which connects it with the main island; its formation is striking as to colour and shape, being of dark grey, and red lava mixed up with ashes, pumice and pozzolana; its most elevated part is a great red and black crag with a flat top rising 1000 feet above the waters of the gulf, which from time to time has been much diminished, as the earthquakes have shaken down large fragments into the sea. Around this crag are the houses of the now ruined town of Skaro, a most extraordinary group, perched one over another where a crevice in the sides of the precipice will admit of it, and in most frightful positions, the outer wall of the buildings being in a line with the perpendicular rocks. I sprung a covey of partridges among this mass of ruins, and was once near falling into a vault full of skulls and other human bones; not a living soul now inhabits this place.

Catholics who had once lived there had moved to the town of Fira.

The Skaros rock is so strongly fractured that one gets the impression that it could drop into the caldera at any moment. For this reason, with the exception of a small church on its north side, is now totally uninhabited. Only a few ruins are left from the formerly numerous dwellings (Fig. 13.7).

Figure 13.7 This photograph of the Skaros promontory was taken in 1993 from nearly the same point as the drawing in Fig. 13.6. After being hit by several severe earthquakes, few traces of the former buildings remain.

The last strong earthquake struck Santorini on the 9th of July, 1956. It was one of the strongest earthquakes in Europe in the twentieth century (Schwarzbach, 1958). In a period of 12.5 minutes, two strong shocks rocked the islands. The epicenter of the first was at the island of Amorgos. The recording stations at Uppsala and Kiruna in Sweden measured a magnitude of 7.7 on the Mercalli–Sieberg scale. The second, which was centered beneath the island of Ios, had a magnitude of 7.2. The first shock had the strength of 9 on the Mercalli–Sieberg scale at Santorini and 8 at Paros. On Therasia, Anaphe, Astypalea, Naxos, Kalymnos, Leros and Patmos it was 7, and in Athens it was 4. The earthquake was felt over an area of 270 000 square kilometers.

The second shock was a little weaker. The Seismological Institute in Athens measured a magnitude of 5 on the Mercalli–Sieberg scale and estimated

Box 13.2 The castle of Skaros

We know from old documents that in the middle of the seventeenth century there were two castles on Skaros, one called Apano Kastro (Upper Castle) and the other Kato Kastro (Lower Castle). In AD 1642 the Roman Catholics were granted 'free space inside the castle for the construction of the bishop's residence using the fallen stones of the walls of the old castle situated above the town'. This was in the same year that the Jesuit priest François Richard came to Santorini. From his account (Richard 1657, p.28) we learn that at that time the older castle on top of the rock of Skaros was in ruins.

The castle is so high that it takes half an hour to climb up to the walls where the doors can be closed when enemies or pirates are seen. In the middle there is a very high rock on top of which there are about 200 houses which are now abandoned and crumbling to ruins. No one is interested in repairing them or in living at such a height. They have given us the whole rock as a place to build our church and our only cost is that of construction. At present there is only one intact church here. People say that a large bell on top of the high rock was rung to warn the inhabitants of the whole island when pirate ships were seen on the sea. Today they use fire as they do on other islands as a warning signal when danger is threatening.

During the eruptions of 1707–1711 the Skaros castle was a place of fear as the Jesuit priest Goree (1712) says:

It was then, that the inhabitants of Santorini, and especially those of the Castle of Scaro, began to be in good earnest afraid. They considered that their Castle was situated upon a Promontory, that was very narrow, and near to the Black Island; and that the time drew near, in which they must expect it either to be blown up into the Air, or overturn'd by some Shock of the Earth. They had continually before their Eyes Fire and Smoak; and this dismal Spectacle made them apprehend, that there might be several Mines of Vitriol and Sulphur in the Island of Santorini, which would soon take Fire; and that therefore the safest way for them was to abandon the Country, and retire to some other Island. And indeed some took this resolution; and there was no other way left to satisfy the rest, but by telling them, that if they would retire further into the Country, they would be safe there; and that if the Castle was in Danger, yet they must necessarily see the Lesser Kammeni first entirely destroyed, not only because it lay between the Castle and the Black Island, but also because it was much nearer to it than to the Castle.

that an area of 180 000 square kilometers was affected. Its impact was especially severe in the towns of Oia and Fira on Thera where almost all the buildings were damaged. Altogether 529 buildings were destroyed, and of these 326 were on Santorini. A total of 1750 were less seriously damaged, and on Santorini and the neighboring islands 53 persons lost their lives. The earthquake also had other serious effects. In Oia the public water cistern was destroyed. Because of the lack of water the village of Agrilia on Therasia had to be evacuated. At the same time the survivors from Mesa Gonia were moved to the fishing village of Kamari on the eastern coast of Thera. The people learned a great deal from this earthquake. Today, all modern buildings are built with reinforcing iron to make them more resistant to strong earthquakes in the future.

Tsunamis

A report from Crete that was written by Philostratos in the first century AD shows how the effects of a tsunami can terrify a population. Earthquakes and a strong recession of the sea were probably related to the appearance of an island (see Box 13.1).

As we have seen in Chapter 12, the eruption of 1650 about seven kilometers off the northeast coast caused a huge tsunami, which flooded the eastern part of Thera. According to eyewitness reports the flood invaded a great part of the land and caused severe damage. Father Goree (1712, pp.357–368) says: 'The

Sea overflowed and destroyed 30000 Perches [about 6 km^2] of Land' and according to Ross (1840) it was two Italian miles. At Kamari and Perissa much loose soil was washed away, exposing pre-Christian ruins at the foot of Mesa Vouno. This catastrophic flood led the people to give up all settlements in low coastal areas. The population was then concentrated in the higher villages like Fira, Oia, Merovigli, Pirgos and Akrotiri.

Tsunamis were also observed at the time of the earthquake of 1956. They were triggered by slumping into an undersea trench near the island of Amorgos. They reached a height of 25 meters on Amorgos, 20 on Astypalea, 5 on Pholegandos, and 4 on Lipsos. On Rhodes and Chios their height was only 0.15 meters (Galanopoulos, 1957). Three great waves were observed within 15 minutes, moving with a velocity of 60 to 90 meters per second. They caused severe damage.

Today, the beach areas at Kamari and Perissa on Thera are densely crowded, but most of the hotels and pensions there are scarcely used in the winter.

Lava flows and ash-falls

The two Kameni islands in the caldera are at present uninhabited and will presumably remain so in the future. The destruction of the summerhouses, churches and harbor on Nea Kameni in the nineteenth century by the lavas and ashes of the Georgios volcano in 1866 to 1870 was a serious lesson (see Chapter 12). There are several geological indications, such as the location of the magma chamber, that a future eruption would most likely take place at the same place. While the effects of lava flows would probably be confined to the area of the caldera, a much wider area could be hit by ash-fall (Box 13.3). Ash can put a heavy weight on the roofs of houses and causes them to collapse. The traditional houses of Santorini have flat roofs for collecting rainwater in cisterns. More recently, however, newly built houses have roofs of reinforced concrete and steel arches that can tolerate minor ash falls without collapsing.

Box 13.3 Were the Argonauts threatened by the Santorini volcano?

The Argonautica of Apollonius Rhodius (*IV.1684–1716*) contains certain elements that could be a reminiscence of volcanic activity experienced by the Argonauts in the sea between Crete and Anaphe. As Luce (1969) has already pointed out, both of these islands lie close to Santorini. The account is an episode in which the giant Talos, 'the man of bronze' (*IV.1638ff.*) hurled blocks (volcanic bombs?) after Jason and his crew and they are covered by an ash-fall ('black chaos coming down from the sky'):

And then straightaway, as they moved swiftly over the great Cretan deep, they were terrified by the night they called 'a pall of darkness'. Neither stars nor moonlight pierced the obscurity. The black chaos was coming down from the sky, or some other form of darkness was rising from the inner most recesses of the earth. They had no way of knowing whether they were voyaging on water or through Hades. Helpless, they could only entrust their safe return to the sea that was carrying them whither it was bearing them. In this extremity Jason prayed to Apollo, and the god guided them with the glint of his bow to Anaphe, east of Thera, where they landed and sacrificed in sunshine once more.

Floating pumice

The eruption of 1650 produced enormous masses of pumice, which not only hindered small boats but also affected people and crops on the neighboring islands. On the island of Ios, waves carried this pumice to an elevation of 20 meters above the normal sea level, and an area more than 120 kilometers in circumference was covered with floating pumice. A manuscript that Ross (1840) found on Thera stated that 'the masses of pumice that drifted toward Crete gave the impression that all of Thera had sunk. The Turkish general in command sent out a vessel to learn what had happened.'

Slumping, landslides, and falling stones

In an area with very high, steep slopes, such as those of the deformed metamorphic rocks in the older massif and the steep inner slopes of the caldera, it is not surprising that earth slides are a common occurrence. The inner walls of the caldera consist of pyroclastic material, some of which is very unstable. Many of the older buildings at the harbor below the town of Fira and at Corfu on Therasia show clear signs of damage by falling rock (Fig. 13.8). The danger is especially serious when earthquakes loosen the rocks. At Cape Therma on the inner side of Thera, for instance, the cave and shelter once used for boats are now shut off by fallen debris (see Fig. 13.3). Roads and paths that go down into the caldera are also endangered. Ludwig Ross (1840) and Karl von Seebach (1867) tell us that people have been killed in such places: 'huge blocks of lava that have been loosened from the ash layers sometimes roll down, especially at night, and kill people or smash into houses at the harbor. Only a few months ago, one of these boulders struck a woman who was on her way up and tore her in two. Twice during the time we were here blocks have fallen on the roofs of the houses.' Even today, one sees crucifixes marking places along the paths where someone has died in this way.

Volcanic gases

During eruptions, the rising magma releases the gases it contains into the atmosphere. Depending on their concentration, these gases can have a toxic effect. If given off during an explosion they can destroy all life in the vicinity of the eruption. The sulfur gases emitted in many volcanic areas are normally only an unpleasant odor. We saw examples of this in the account of visits to Nea Kameni in 1707 (Chapter 12). If they are in stronger concentrations, however, they can have serious consequences.

The sulfur gas, hydrogen sulfide, is easily detected because it has a strong odor of rotten eggs and a yellowish color from sulfur, but carbon dioxide or carbon monoxide is a different matter. These gases have neither color nor odor and they can overcome a person without warning. Emissions of carbon dioxide, known as mofettes, are common in volcanic craters and areas of older lavas that are still cooling. Because it is heavier than the air, carbon dioxide flows directly over the ground surface, and accumulates in depressions. Dogs, sheep, and other small animals are sometimes killed while taller humans feel no effect. There are places in the Eifel region of Germany where natural carbon dioxide of this kind is a well-known danger. The gas may even collect in the basements of houses, and gravediggers have been known to die while digging graves. In recent years 1700 people in a single village died from emissions of carbon dioxide at Lake Nyos in Cameroon, and 156 people died from similar causes in Java.

Deaths from gas emissions have also occurred on Santorini. We know from contemporary records that in 1650 great amounts of sulfur gases were released during an eruption outside the caldera seven kilometers from Cape Kolumbo. The odor of sulfur was strong and irritating. Articles of silver turned black, even when stored in closed boxes. We have only indirect knowledge of carbon dioxide. Ludwig Ross (1840) recounts events described in a Greek manuscript he found on Thera (see Box 13.4).

Monitoring and predicting natural hazards

Gravimetric and geomagnetic measurements

Measurements of small variations of the local gravitational field are an important tool for investigating the internal structure of volcanoes. A survey of this kind is being conducted by IGME from Athens in collaboration with the Geological and Geophysical Institute of the University of Athens. A series of leveling points has been established for this purpose on Santorini, so that measurement can be made at regular intervals to detect changes in the state of the volcano. The method

Figure 13.8 The steep caldera wall formed by the Minoan eruption continues to be unstable 3600 years later. From time to time, large boulders are loosened from the wall and tumble down with a terrific noise. In the spring of 1993, about 100 square meters of the rock wall between Cape Plaka and Cape Thermia fell with such a roar that people in Megalochori were awaken. The large rock seen in this photograph of the beach on Therasia broke away from a height of about 150 meters, rolled down the slope, and demolished two houses at the harbor of Corfu.

Box 13.4 Poisonous gas during the eruption of AD 1650

When the eruption of the Kolumbo volcano in 1650 took place outside the caldera, poisonous volcanic gases invaded Santorini. Ross (1840) has translated a manuscript that records this event. It speaks of sulfur gases, but there must also have been carbon dioxide.

A heavy cloud filled with deadly gases rested on the sea and on the shallow coast, for, as the writer of the manuscript reports, several persons died when they went too close to the coast to collect fish washed up on the shore. Numerous sheep and other livestock, including even grouse and other wild birds, were found dead in the fields. On October the 2nd two Theran boats came from Amorgos. The first reached Apanomeria [today Oia] safely even though the sailors had been in danger of asphyxiation. They survived by putting wine-soaked cloths over their noses and breathing through them. The other boat, however, got trapped in the masses of floating pumice, and after a few days, when the inhabitants of Ios brought it to shore, they found the sailors asphyxiated and their bodies swollen.

is based on the principle that rising magma changes the distribution of mass and the local gravity field. Very small changes of elevation can also be detected. The technique has been used successfully on Vesuvius and Etna and on Kilauea in Hawaii. Records acquired on Santorini since 1984 show that some of the measuring points on Thera have been slightly elevated while others were slightly depressed (Lagios *et al.*, 1990a). This probably indicates that magma is still active in the reservoir under the volcano.

Measurements of the geomagnetic field are a similar geophysical method used for forecasting eruptions. This makes it possible to locate the approximate position of the magma chamber and detect changes that might be a sign of unrest.

Seismic monitoring

Of all the methods available for monitoring volcanoes, recording seismic activity is by far the most useful. Shallow earthquakes that are easily detected precede every volcanic eruption. The seismographic record is also useful for studying the subsurface structure.

In the case of Santorini, for example, earthquakes have been found to be concentrated at rather shallow depths (less than 5 kilometers). The epicenters are restricted to two areas, the Kameni islands and the area around Kolumbo volcano, both places where eruptions have occurred in historic times (Delibasis *et al.*, 1990).

Tectonic lines

Forecasting volcanic eruptions requires a thorough knowledge of the history and tectonic setting of the volcano. Eruptions occur mainly in recognizable zones of weakness that can usually be seen on aerial photographs, measurements in the field, and the distribution of earthquakes. By compiling a record of past activity, it is possible to detect trends in the distribution of eruptions (Fig. 13.9) and how the volcano has behaved in the past. With this information one can detect zones where eruptions are likely to occur in the future. For example during the eruptions of 1707–1711, 1866–1870 and 1925–1928 all changes in eruptive activity proceeded in the same direction, namely northeast–southwest. The inhabitants of the Skaros rock had already realized this in 1707. When the eruption in that year started in the caldera they understood that their houses were directly in the 'line of fire' along which the last two eruptions had occurred. The vents of the present Kameni islands define one such zone, and another is marked by the eruptive centers extending from the central vent of Megalo Vouno to the Kolumbo volcano. Future eruptions are therefore likely to follow these two zones of weakness.

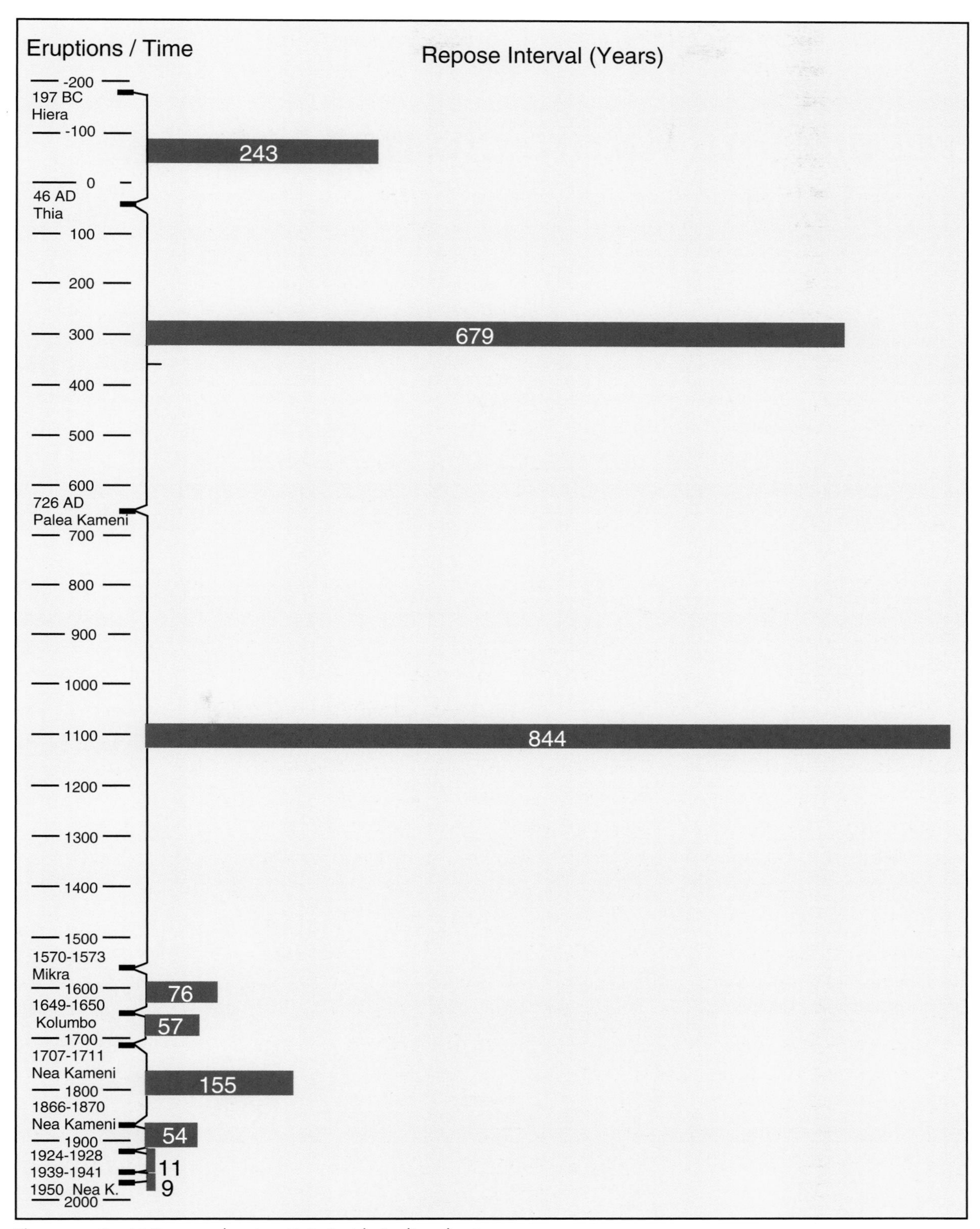

Figure 13.9 Post-Minoan volcanic activity inside (red) and outside (orange) the caldera. Horizontal bars show the repose interval between eruptions.

14

Present and future changes of the island

The forces of nature and man are changing the appearance of the volcanic islands. The unrelenting erosive forces of the sea, wind, and rain destroy, with time, what the volcano has deposited during the relatively brief time of an eruption. The various types of rocks making up the islands react differently to the effects of this erosion.

Erosion and its attendant effects on the landscape are among humankind's oldest experiences. The ancient Greeks understood this power of nature. In his well-known work *Timaios,* Plato used the image of slow but steady changes of the landscape as a symbol for long spans of time. Santorini is exposed to these forces, which, even today, are changing the appear-

Figure 14.0 Extremely good visual phenomena can result from certain weather conditions, such as those during the hurricanes in the first days of January 1980. The meteorologist Wilski, who worked together with Hiller von Gaertringen on the excavation of ancient Thera, reported that one could even see Crete from Santorini, a distance of 120 kilometers. Under such conditions up to seventeen islands may be seen in the vicinity of Santorini. In this photograph the island of Ios can be seen on the horizon to the left. In the caldera the strong waves breaking on the shore have removed pumice from the coast; this forms a garland-like chain on the water.

ance of the islands. The present landscape is a remnant of the ring-island created by thousands of years of volcanism, but its appearance is already quite different from what it was just after the Minoan eruption. Weathering and erosion have wrought clear changes to the landscape. The relatively resistant rocks of the metamorphic basement respond differently to the forces of erosion than does the loose volcanic ash, so that, with time, the rocks are sculptured into new forms. Even before the first eruptions, the Cycladic metamorphic rocks were subjected to prolonged erosion. Following the Tertiary deformation responsible for their intense folding they were buried beneath a thick blanket of volcanic rocks, but only a relatively short time elapsed before much of the loose volcanic debris was stripped away exposing the metamorphic basement. As erosion continues, part of the debris derived from the non-volcanic period accumulates at the foot of the massif of Profitis Elias. Thus the phyllites that are partly exposed in the caldera wall at Athinios, Cape Plaka and Cape Therma are protected from erosion by a mantle of volcanic ash and lava that covers them.

The height of the old basement series at Profitis Elias (565 meters) is due to earlier uplifting. Its steep slopes are especially vulnerable to erosion along ancient structural features, such as the graben at Selada. During the heavy winter rains, this graben acts like a drainage channel for floods coming down the slopes and carrying sediments eroded from the underlying rocks. The landing place of Ancient Oia, used by the Greeks and Romans from about 800 BC to around 800 AD, is a product of this process. Hiller Von Gaertringen found this landing place at the beginning of the twentieth century near the southernmost edge of the present town of Kamari.

The erosion channels on Profitis Elias remind me of an incident that happened around 1984. In connection with the water that is so precious on Santorini, I was asked by Lefteris Sigalas, hotel owner and brother of the archaeologist Ch. Sigalas from Kamari, where in this area I would drill for water. Although I had never before been involved in such matters, I found the problem interesting and investigated the area on the flood plain below Profitis Elias. I noticed that the old erosional channels that run down the slope and carry water during the winter rains passed under the cover of Minoan pumice blanketing the plain of Kamari. One can see from the geological conditions in the nearby hills that the metamorphic basement rocks extend beneath this plain and that the water running down slope might continue to follow these old channels where they are covered by pumice. Together with Lefteris, I went into the field and discussed with him the geological conditions, explained the role of the old erosional channels, and suggested that 'one should drill for water where the erosional channels coming down the mountain meet the plain'. He followed my advice, and in his opinion the well proved to be one of the best on Santorini. Beginner's luck is an old term for this sort of success, but this experience convinced the people on Santorini that I was able to find water. I have declined further requests of this kind, however, because it would not be right for me to compete with my Greek colleagues.

The relatively young part of Santorini is subjected to strong erosional forces (Fig. 14.1). Sea, rain and wind can easily erode the soft, pyroclastic rocks. On the cliffs, erosion has produced remarkable patterns and forms in the relatively soft pumice (Figs. 14.2 and 14.3). Only where lava mantles the soft layers do the latter escape rapid erosion. This is particularly important on the Kameni Islands.

Erosion is not normally strong in mild weather, but under extreme conditions, such as hurricanes (Fig. 14.0), the air is filled with so much pumice that it resembles an eruption cloud. As in a desert sandstorm, one must seek protection from the sandblasting of the wind (Figs. 14.4 and 14.5). It is not surprising that we find so many ancient altars that were devoted to the wind god, for strong storms and heavy rainfalls are not uncommon in the winter on Santorini (Fig. 14.6).

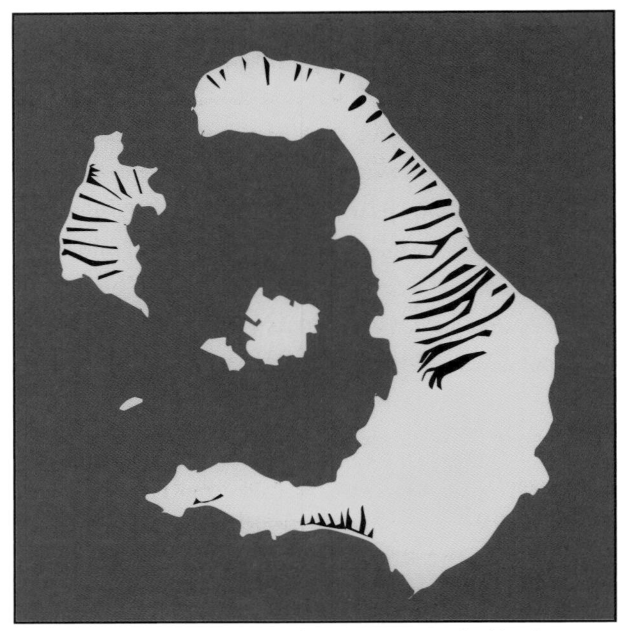

Figure 14.1 The topography of Santorini is marked by deep erosional channels that have cut into the surface layer of soft pumice. During winter rains, the rainwater can form torrents, which the Greeks call potamos. Many villages are sheltered in these erosion channels, where they are protected from the strong winter storms.

Erosion due to flooding

Loose pyroclastic material can also be reworked by other means. When the roof of the magma chamber collapsed during the Minoan eruption huge masses of water must have rushed into the depths and caused tsunamis. The rebounding water could have spilled over the caldera rim and carried away loose tuff from the slopes while leaving the heavier lithic blocks behind. The eruption of 1650 showed how this may have happened. At that time great parts of the eastern flank of Thera were stripped away, exposing long-buried buildings at Kamari and Perissa (see Chapter 12).

One can see on aerial photographs of Thera a system of radial erosional channels in the flat area east of the town of Akrotiri. These probably originated during the last phase of the Minoan eruption when large waves or tsunamis washed over the caldera rim.

Figure 14.2 Erosion has also left its marks on the steep caldera walls, where caves and grottoes have been cut into the soft volcanic rocks. From ancient times the inhabitants of Santorini have used these places for their houses and churches. The Christos church at Thermia, shown here, is on the inner side of the caldera.

Figure 14.3 Erosion by wind and water can produce bizarre formations in the soft pumice of the Minoan eruption. This pattern was visible on the east hills of Thera in 1990. Wind erosion produces these strange patterns by preferentially removing soft materials and leaving others that are more resistant. The relief reveals pyroclastic flow structures. The wall has a height of about 10 meters.

Figure 14.4 Honeycomb-like patterns are often seen in the rocks of the coastal area. This soft lava block shows the characteristic form of what is called 'tafoni weathering'.

Figure 14.5 The steep walls of the caldera are not yet stable: slumping of the light pumice layers is quite common. Fresh detrital pumice is sometimes so inflated with air that one can sink to the hips in the loose powdery material. It is also easily mobilized and can flow like a fluid. It can also easily be blown by the wind and cause sandblasting.

Changes on the islands due to human activity

Santorini is a very popular place for tourism. Although it is now visited throughout the year by thousands of people, it has not always been like this. In former years ships were the only connection to the outer world, and tourists were seen only during the summer season. Since the middle of the 1970s, however, air connection was brought to Thera and direct air travel was possible from several European cities, especially during the summer. This has led to an enormous boom in tourism. At present, Santorini has a capacity of about 80 000 beds. The island group has many attractions, such as the active volcano with its fumaroles and solfatara. There are also the sunny skies, the beaches, and of course the exciting archaeological discoveries. Santorini is 'in'.

Figure 14.6 Trees grow mainly in erosional channels where they find protection from the wind. On slopes they adjust their growth form to the prevailing direction of the wind.

But the story has another side that is not so pleasant. Many people that have known Santorini and have friends on the islands realize that the hospitable, amiable inhabitants of the island are motivated more and more by profit. The tendency, unfortunately, is visible everywhere. Quarrels about rights of paths, boundaries, and fields are daily occurrences. Lawyers and courts have much to do. The huge boom of tourism is changing not only the inhabitants of the island but the landscape as well. The disposal of garbage does not function very well. The Kameni islands, one of the attractions of Santorini that are visited by thousands of tourists every year, suffer from this problem. Empty water bottles and cans litter the ground, and the wind spreads debris everywhere. The fascinating active volcanoes should be a natural preserve. The same is also true of places on the caldera walls where archaeological sites have been found and where one can study unusual geological phenomena. These geological and archaeological sites could become new attractions for Santorini with guides and informative signs. They should be given higher priority than the building of hotels and garbage dumps.

In short, many of Santorini's natural beauties are threatened by the expanding tourism, and even though the inhabitants say they lack money to protect the landscape and antiquities, it is still possible to act and preserve the future of Calliste.

Strange structures on the sea floor

In 1967, when James Mavor and Angelos Galanopoulos were looking for traces of Atlantis on Thera they investigated the shallow waters in and around the caldera. Mavor and his co-workers made depth soundings and underwater surveys at Kamari, Perissa, and Cape Exomytis. It was a note written by a British naval officer named Leycester that had attracted them to this place. In 1848 Leycester had published a navigation chart of Santorini and reported ruins that were said to occur at several places in water ten to thirty meters deep. He reported various underwater structures, including even the remains of chimneys, and claimed to have found a large part of a harbor, which he showed on his map. Ludwig Ross (1840) had already reported similar phenomena at Exomytis. In his view, the underwater ruins could be remains of the city of Eleusis, the town that, according to Ptolemy (2nd century BC), should have been situated on Thera.

According to Ross, when the sea is calm one can see from the nearby hills the walls of buildings under the sea (see Box 14.1).

At the time of their investigation, Mavor and Galanopoulos already knew that the so-called harbor piers at Exomytis were natural formations. Nature makes a stone formation in the sea that is quite similar to walls of buildings. This is called 'beach rock' because it is formed close to the seashore. In places near Kamari and Perissa this beach rock had produced terrace-like 'concrete platforms' that can be very unpleasant for bathers when they are covered by a slippery mass of algae. The findings at Cape Exomytis have a natural explanation and this enigma can now be considered solved. Mavor and Galanopoulos found that the 'natural harbor' at Exomytis would not have been very suitable, because it was exposed to winds and currents. This observation in itself suggested that the feature had a natural origin. It is interesting that Mavor also reported a shallow underwater ring-like structure with a diameter of one meter that in his opinion was not made by nature but by man. Mavor took this as further evidence for Atlantis.

The formation of beach rock is an on-going process that, even today, is gradually changing the coastal regions (Fig. 14.7). It occurs where groundwater flows over or through limestone and dissolves carbonate ions from the rock. When the groundwater meets warm sea water near the beach it precipitates the carbonate between the sand grains or pebbles and forms a concrete-like mass. Thus, the occurrence of beach rock is a sign that one can find fresh water nearby. On Santorini we also find trees and other forms of vegetation, such as bushes, in these areas.

Box 14.1 Is there a city of Eleusis at Cape Exomytis?

Observations by the archaeologist Ludwig Ross (1840 *Inselreisen*, page 58):

Leaving the village of Emporia on our right, we rode across the coastal plain to the hills of Exomytis, which project, from the main mountains in a south-western direction. Ptolemy mentioned Eleusis as the second of the main villages of the island. It was probably situated close to the south side of the foothills. Its ruins may have sunk into the sea in the dark mediaeval centuries when the town was struck by one of the earthquakes that are so common in this area. During calm weather, one can see under-water ruins, and the form of the pier is clearly visible.

Lieutenant Leycester, who visited Santorini in 1848, also reported in his paper:

Some Account of the Volcanic Group of Santorin or Thera, once called Callistê, or the Most Beautiful' on the underwater ruins at Cape Exomytis. In a detailed map, drawn by Captain Thomas Graves, he shows the position of the 'ancient moles'.

On pages 29–30 he says:

'I was amused one day by my Greek interpreter bringing me information that he had received from the inhabitants of the country, that there were houses under water here, and that in calm weather their chimneys were to be seen. I wonder whether, in sounding, our lead ever intruded into the kitchens of the inhabitants of ancient Eleusis!'

Volcanoes are unpredictable

The steadily moving lithospheric plates of the earth create tectonic stresses and regions of active volcanoes, such as those of the southern Aegean volcanic arc. That means that earthquakes, eruptions, and tsunamis will continue to characterize this area in the future (Fritzalas, 1988). Nothing can stop this dynamic force. The geologic development of Santorini in the course of time is the best proof of this (Fig. 14.8). As long as the tectonic stresses that build up with time are relieved by small, frequent movements, the reaction in the earth's crust will be correspondingly weak: minor earthquakes and smaller but more frequent eruptions will be the result. If there are eruptions in the near future similar to those that occurred on Nea Kameni in this century, they will be relatively harm-

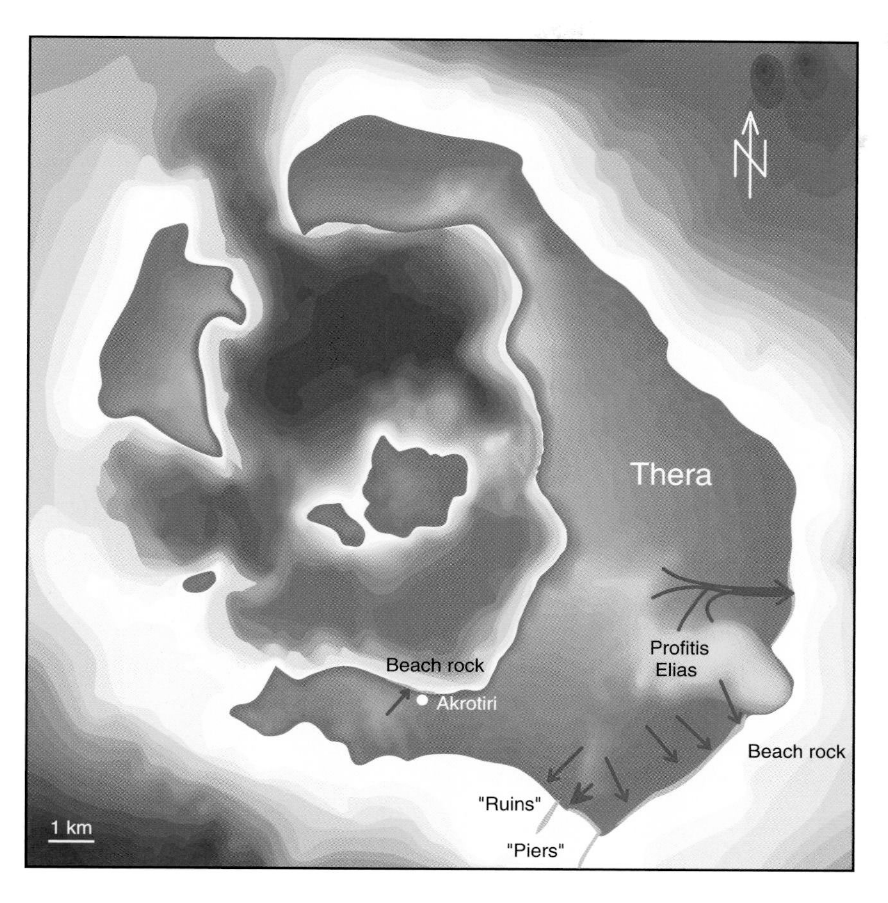

Figure 14.7 Localities where beach rock is found today on Santorini (red). Groundwater (blue arrows) running down the slopes of Profitis Elias and Platinamos contains carbonates dissolved from the limestone and marble. When it meets the warm sea water, the carbonate is precipitated and cements the sand and gravel near the shore forming a concrete-like platform. In shallow water, pieces of broken beach rock can resemble underwater ruins and piers.

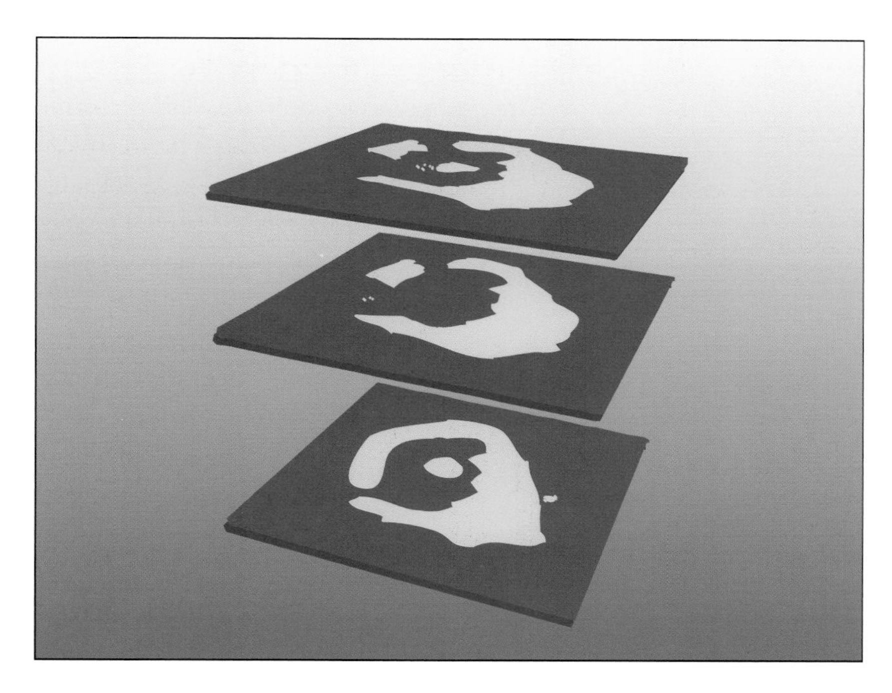

Figure 14.8 The changes on Santorini since the Bronze Age are the result of the irresistible motion of lithospheric plates that is the ultimate cause of volcanism. In the future, this picture will continue to change. The lower part of the diagram shows the nearly closed ring-island with the Pre-Kameni island in the middle and the Monolithos rock offshore, as it looked before the Minoan eruption. The middle diagram shows the shape of Santorini after the Minoan eruption, with the islands of Thera, Therasia, and Aspronisi. The present form of Santorini is shown in the upper part where the two central islands of Nea and Palea Kameni have formed in the last 2000 years.

less. As the volcanologist Alexander R. McBirney says: 'The greatest danger will not be the flow of lava coming down the volcano but the flow of tourists coming to see it'.

If, however, the tectonic stress accumulates over a longer period the impending danger of strong earthquakes and violent eruptions is greater. Because calc-alkaline magmas of the kind found in the southern Aegean arc are exceptionally gas-rich and explosive, they have the potential for strong explosions. No one knows, of course, when there will be another event with the magnitude of the Minoan eruption. 'Volcanoes are unpredictable' as the volcanologist Maurice Kraft once said. He proved this when, together with his wife Katja, he was killed while observing the eruption of Unzen, Japan, in June 1991. On Santorini the possibilities of further eruptions are equally uncertain, and we cannot be sure they will follow the same zones of weakness inside and outside the caldera, for they may break out somewhere new. The fire in the sea is not yet extinguished and it is only a question of time until it will come to life again.

Appendix 1

Plato's dialogs, Critias and Timaeus

CRITIAS: Then listen, Socrates, to a tale which, though strange, is certainly true, having been attested by Solon, who was the wisest of the seven sages. He was a relative and a dear friend of my great-grandfather, Dropides, as he himself says in many passages o his poems; and he told the story to Critias, my grandfather, who remembered and repeated it to us. There were of old, he said, great and marvellous actions of the Athenian city, which have passed into oblivion through lapse of time and the destruction of mankind, and one in particular, greater than all the rest. This we will now rehearse. It will be a fitting monument of our gratitude to you, and a hymn of praise true and worthy of the goddess, on this her day of festival.

SOCRATES: Very good. And what is this ancient famous action of the Athenians, which Critias declared, on the authority of Solon, to be not a mere legend, but an actual fact?

CRITIAS: I will tell an old-world story which I heard from an aged man; for Critias, at the time of telling it, was as he said, nearly ninety years of age, and I was about ten. Now the day was that day of the Apaturia which is called the Registration of Youth, at which, according to custom, our parents gave prizes for recitations, and the poems of several poets were recited by us boys, and many of us sang the poems of Solon, which at that time had not gone out of fashion.

One of our tribe, either because he thought so or to please Critias, said that in his judgment Solon was not only the wisest of men, but also the noblest of poets. The old man, as I very well remember, brightened up at hearing this and said, smiling: Yes, Amynander, if Solon had only, like other poets, made poetry the business of his life, and had completed the tale which he brought with him from Egypt, and had not been compelled, by reason of the factions and troubles which he found stirring in his own country when he came home, to attend to other matters, in my opinion he would have been as famous as Homer or Hesiod, or any poet.

And what was the tale about, Critias? said Amynander.

About the greatest action which the Athenians ever did, and which ought to have been the most famous, but, through the lapse of time and the destruction of the actors, it has not come down to us.

Tell us, said the other, the whole story, and how and from whom Solon heard this veritable tradition. He replied:

In the Egyptian Delta, at the head of which the river Nile divides, there is a certain district which is called the district of Sais, and the great city of the district is also called Sais, and is the city from which King Amasis came. The citizens have a deity for their foundress; she is called in the Egyptian tongue Neith, and is asserted by them to be the same whom the Hellenes call Athene; they are great lovers of the Athenians, and say that they are in some way related to them.

To this city came Solon, and was received there with great honour; he asked the priest who were most skilful in such matters, about antiquity, and made the discovery that neither he nor any other Hellene knew anything worth mentioning about the times of old. On one occasion, wishing to draw them on to speak of antiquity, he began to tell about the most ancient things in our part of the world – about Phoroneus, who is called 'the first man,' and about Niobe; and after the Deluge, of the survival of Deucalion and Pyrrha; and he traced the genealogy of their descendants, and reckoning up the dates, tried to compute how many years ago the events of which he was speaking happened.

Thereupon one of the priests, who was of a very great age, said: O Solon, Solon, you Hellenes are never anything but children, and there is not an old man among you. Solon in return asked him what he meant. I mean to say, he replied, that in mind you are all young; there is no old opinion handed down among you by ancient tradition, nor any science which is hoary with age, And I will tell you why.

There have been, and will be again, many destructions of mankind arising out of many causes; the greatest have been brought about by the agencies of fire and water, and other lesser ones by innumerable other causes. There is a story, which even you have preserved, that once upon a time Paethon, the son of Helios, having yoked the steeds in his father's chariot, because he was not able to drive them in the path of his father, burnt up all that was upon the earth, and was himself destroyed by a thunderbolt. Now this has the form of a myth, but really signifies a declination of the bodies moving in the heavens around the earth, and a great conflagration of things upon the earth, which recurs after long intervals; at such times those who live upon the mountains and in dry and lofty places are more liable to destruction than those who dwell by rivers or on the seashore. And from this calamity the Nile, who is our never-failing saviour, delivers and preserves us.

When, on the other hand, the gods purge the earth with a deluge of water, the survivor in your country are herdsmen and shepherds who dwell on the mountains, but those who, like you, live in cities are carried by the rivers into the sea. Whereas in this land, neither then nor at any other time, does the water come down from above on the fields having always a tendency to come up from below; for which reason the traditions preserved here are the most ancient. The fact is, that wherever the extremity of winter frost or of summer does not prevent, mankind exist, sometimes in greater, sometimes in lesser numbers. And whatever happened either in your country or in ours, or in any other way remarkable, they have all been written down by us of old, and are preserved in our temples.

Whereas just when you and other nations are beginning to be provided with letters and the other requisites of civilized life, after the usual interval, the stream from heaven, like a pestilence, comes pouring down, and leaves only those of you who are destitute of letters and education; and so you have to begin all over again like children, and know nothing of what happened in ancient times, either among us or among yourselves. As for those genealogies of yours which you just now recounted to us, Solon, they are no better than the tales of children.

In the first place you remember a single deluge only, but there were many previous ones; in the next place, you do not know that there formerly dwelt in your land the fairest and noblest race of men which ever lived, and that you and your whole city are descended from a small seed or remnant of them which survived. And this was unknown to you, because, for many generations, the survivors of that destruction died, leaving no written word. For there was a time, Solon, before the great deluge of all, when the city which now is Athens was first in war and in every way the best governed of all cities, is said to have performed the noblest deeds and to have had the fairest constitution of any of which tradition tells, under the face of heaven.

Solon marvelled at his words, and earnestly requested the priests to inform him exactly and in order about these former citizens. You are welcome to hear about them, Solon, said the priest, both for your own sake and for that of your city, and above all, for the sake of the goddess who is the common patron and parent and educator of both our cities. She founded your city a thousand years before ours, receiving from the Earth and Hephaestus the seed of your race, and afterwards she founded ours, of which the constitution is recorded in our sacred registers to be eight thousand years old.

As touching your citizens of nine thousand years ago, I will briefly inform you of their laws and of their most famous action; the exact particulars of the whole we will hereafter go through at our leisure in the sacred registers themselves. If you compare these very laws with ours you will find that many of ours are the counterpart of yours as they were in the olden time.

In the first place, there is the caste of priests, which is separated from all the others; next, there are the artificers, who ply their several crafts by themselves and do not intermix; and also there is the class of shepherds and of hunters, as well as that of husbandmen; and you will observe, too, that the warriors in Egypt are distinct from all the other classes, and are commanded by the law to devote themselves solely to military pursuits; moreover, the weapons which they carry are shields and spears, a style of equipment which the goddess taught of Asiatics first to us, as in your part of the world first to you.

Then as to wisdom, do you observe how our law from the very first made a study of the whole order of things, extending even to prophecy and medicine which gives health, out of these divine elements deriving what was needful for human life, and adding every sort of knowledge which was akin to them. All this order and arrangement the goddess first imparted to you when establishing your city; and she chose the spot of earth in which you were born, because she saw that the happy temperament of the seasons in that land would produce the wisest of men. Wherefore the goddess, who was a lover both of war and of wisdom, selected and first of all settled that spot which was the most likely to produce men likest herself. And there you dwelt, having such laws as these and still better ones, and excelled all mankind in all virtue, as became the children and disciples of the gods.

Many great and wonderful deeds are recorded of your state in our histories. But one of them exceeds all the rest in greatness and valour. For these histories tell of a mighty power which unprovoked made an expedition against the whole of Europe and Asia, and to which your city put an end. This power came forth out of the Atlantic Ocean, for in those days the Atlantic was navigable; and there was an island situated in front of the straits which are by you called the Pillars of Heracles; the island was larger than Libya and Asia put together, and was the way to other islands, and from these you might pass to the whole of the opposite continent which surrounded the true ocean; for this sea which is within the Straits of Heracles is only a harbour, having a narrow entrance, but that other is a real sea, and the surrounding land may be most truly called a boundless continent.

Now in this island of Atlantis there was a great and wonderful empire which had rule over the whole island and several others, and over parts of the continent, and, furthermore, the men of Atlantis had subjected the parts of Libya within the columns of Heracles as far as Egypt, and of Europe as far as Tyrrhenia. This vast power, gathered into one, endeavoured to subdue at a blow our country and yours and the whole of the region within the straits; and then, Solon, your country shone forth, in the excellence of her virtue and strength, among all mankind. She was pre-eminent in courage and military skill, and was the leader of the Hellenes. And when the rest fell off from her, being compelled to stand alone, after having undergone the very extremity of danger, she defeated and triumphed over the invaders, and preserved from slavery those who were not yet subjugated, and generously liberated all the rest of us who dwell within the pillars.

But afterwards there occurred violent earthquakes

and floods; and in a single day and night of misfortune all your warlike men in a body sank into the earth, and the island of Atlantis in like manner disappeared in the depths of the sea. For which reason the sea in those parts is impassable and impenetrable, because there is a shoal of mud in the way; and this was caused by the subsidence of the island.

I have told you briefly, Socrates, what the aged Critias heard from Solon and related to us. And when you were speaking yesterday about your city and citizens, the tale which I have just been repeating to you came into my mind, and I remarked with astonishment how, by some mysterious coincidence, you agreed in almost every particular with the narrative of Solon; but I did not like to speak at the moment. For a long time had elapsed, and I had forgotten too much; I thought that I must first of all run over the narrative in my own mind, and then I would speak.

And so I readily assented to your request yesterday, considering that in all such cases the chief difficulty is to find a tale suitable to our purpose, and that with such a tale we should be fairly well provided. And therefore, as Hermocrates has told you, on my way home yesterday I at once communicated the tale to my companions as I remembered it and after I left them, during the night by thinking I recovered nearly the whole it. Truly as is often said, the lessons of our childhood make wonderful impression on our memories; for I am not sure that I could remember all the discourse of yesterday, but should be much surprised if I forgot any of these things which I have heard very long ago. I listened at the time with childlike interest to the old man's narrative; he was very ready to teach me, and I asked him again and again to repeat his words, so that like an indelible picture they were branded into my mind.

As soon as the day broke, I rehearsed them as he spoke them to my companions, that they, as well as myself, might have something to say. And now, Socrates, to make an end my preface, I am ready to tell you the whole tale. I will give you not only the general heads, but the particulars, as they were told to me.

The city and citizens, which you yesterday described to us in fiction, we will now transfer to the world of reality. It shall be the ancient city of Athens, and we will suppose that the citizens whom you imagined, were our veritable ancestors, of whom the priest spoke; they will perfectly harmonise, and there will be no inconsistency in saying that the citizens of your republic are these ancient Athenians. Let us divide the subject among us, and all endeavour according to our ability gracefully to execute the task which you have imposed upon us. Consider then, Socrates, if this narrative is suited to the purpose, or whether we should seek for some other instead.

HERM. And in truth, Socrates, you are giving me the same warning as Critias. But men of faint heart never yet set up a trophy, Critias; wherefore you must go forward to your discoursing manfully, and, invoking the aid of Paion and the Muses, exhibit and celebrate the excellence of your ancient citizens.

CRIT. You, my dear Hermocrates, are posted in the last rank, with another man before you, so you are still courageous. But experience of our task will of itself speedily enlighten you as to its character. However, I must trust to your consolation and encouragement, and in addition to the gods you mentioned I must call upon all the rest and especially upon Mnemosynê. For practically all the most important part of our speech depends upon this goddess; for if I can sufficiently remember and report the tale once told by the priests and brought hither by Solon, I am wellnigh convinced that I shall appear to the present audience to have fulfilled my task adequately. This, then, I must at once proceed to do, and procrastinate no longer.

Now first of all we must recall the fact that 9000 is the sum of years since the war occurred, as is recorded, between the dwellers beyond the pillars of Heracles and all that dwelt within them; which war we have now to relate in detail. It was stated that this city of ours was in command of the one side and fought through the whole of the war, and in command of the other side were the kings of the island of Atlantis, which we said was an island larger than Libya and Asia once upon a time, but now lies sunk by earthquakes and has created a barrier of impassable mud which prevents those who are sailing out from here to the ocean beyond from proceeding further. Now as regards the numerous barbaric tribes and all the Hellenic nations that then existed, the sequel of our story, when it is, as it were, unrolled, will disclose what happened in each locality; but the facts about the Athenians of that age and the enemies with whom they fought we must necessarily describe first, at the outset, – the military power, that is to say, of each and their forms of government. And of these two we must give the priority in our account to the state of Athens.

Once upon a time the gods were taking over by lot the whole earth according to its regions, – not according to the results of strife: for it would not be reasonable to suppose that the gods were ignorant of their own several rights, nor yet that they attempted to obtain for themselves by means of strife a possession to which others, as they knew, had a better claim. So by just allotments they received each one his own, and they settled their countries; and when they had thus settled them, they reared us up, even as herdsmen rear their flocks, to be their cattle and nurslings; only it was not our bodies that they constrained by bodily force, like shepherds guiding their flocks with stroke of staff, but they directed from the stern where the living creature is easiest to turn about, laying hold on the soul by persuasion, as by a rudder, according to their own disposition; and thus they drove and steered all the mortal kind. Now in other regions others of the gods had their allotments and ordered the affairs, but in as much as Hephaestus and Athena were of a like nature, being born of the same father, and agreeing, moreover, in their love of wisdom and of craftsmanship, they both took for their joint portion this land of ours as being naturally congenial and adapted for virtue and for wisdom, and therein they planted as native to the soil men of virtue and ordained to their mind the mode of government. And of these citizens the names are preserved, but their works have vanished owing to the repeated destruction of their successors and the length of the intervening periods. For, as was said before, the stock that survived on each occasion was a remnant of unlettered mountaineers which had heard the names only of the rulers, and but little besides of their works. So though they gladly passed on these names to their descendants, concerning the mighty deeds and the laws of their predecessors they had no knowledge, save for some invariably obscure reports; and since, moreover, they and their children for many generations were themselves in want of the necessaries of life, their attention was given to their own needs and all their talk was about them; and in eonsequenee they paid no regard to the happenings of bygone ages. For legendary lore and the investigation of antiquity are visitants that come to cities in company with leisure, when they see that men are already furnished with the necessaries of life, and not before.

In this way, then, the names of the ancients, without their works, have been preserved. And for evidence of what I say I point to the statement of Solon, that the Egyptian priests, in describing the war of that period, mentioned most of those names – such as those of Cecrops and Erechtheus and Erichthonius and Erysichthon and most of the other narmes which are recorded of the various heroes before Theseus – and in like manner also the names of the women. Moreover, the habit and figure of the goddess indicate that in the case of all animals, male and female, that herd together, every species is naturally capable of practising as a whole and in common its own proper excellence.

Now at that time there dwelt in this country not only the other classes of the citizens who were occupied in the handicrafts and in the raising of food from the soil, but also the military class, which had been separated off at the commencement by divine heroes and dwelt apart. It was supplied with all that was required for its sustenance and training, and none of its members possessed any private property, but they regarded all they had as the common property of all; and from the rest of the citizens they claimed to receive nothing beyond a sufficiency of sustenance and they practised all those pursuits which were mentioned yesterday, in the description of our proposed 'Guardians'. Moreover, what was related about our country was plausible and true, namely, that, in the first place, it had its boundaries at that time marked off by the Isthmus, and on the inland side reaching to the heights of Cithaeron and Parnes, and that the boundaries ran down with Oropia on the right, and on the seaward side they shut off the Asopus on the left; and that all other lands were surpassed by ours in goodness of soil, so that it was actually able at that period to support a large host which was exempt from the labours of husbandry. And of its goodness a strong proof is this: what is now left of our soil rivals any other in being all productive and abundant in crops and rich in pasturage for all kinds of cattle; and at that period, in addition to their fine quality it produced these things in vast quantity How, then, is this statement plausible, and what residue of the land then existing serves to confirm its truth? The whole of the land lies like a promontory jutting out from the rest of the continent far into the sea; and all the cup of the sea round about it is, as it happens, of a great depth. Consequently, since many great convulsions took place during the 9000 years – for such was the number of years from that time to this – the soil which has kept breaking away from the high lands during these ages and these disasters, forms no pile of sediment worth mentioning, as in other regions, but keeps sliding away ceaselessly and disappearing in the deep. And, just as happens in small islands, what now remains compared with what then existed is like the skeleton of a sick man, all the fat and soft earth having wasted away, and only the bare framework of the land being left. But at that epoch the country was unimpaired, and for its mountains it had high arable hills, and in place of the 'moorlands', as they are now called, it contained plains full of rich soil; and it had much forest-land in its mountains, of which there are visible signs even to this day; for there are some mountains which now have nothing but food for bees, but they had trees no very long time ago, and the rafters from those felled there to roof the largest buildings are still sound. And besides, there were many lofty trees of cultivated species; and it produced boundless pasturage for flocks. Moreover, it was enriched by the yearly rains from Zeus, which were not lost to it, as now, by flowing from the bare land into the sea; but the soil it had was deep, and therein it received the water, storing it up in the retentive loamy soil; and by drawing off into the hollows from the heights the water that was there absorbed, it provided all the various districts with abundant supplies of spring-waters and streams, whereof the shrines which still remain even now, at the spots where the fountains formerly existed, are signs which testify that our present description of the land is true.

Such, then, was the natural condition of the rest of the country, and it was ornamented as you would expect from genuine husbandmen who made husbandry their sole task and who were also men of taste and of native talent, and possessed of most excellent land and a great abundance of water, and also, above the land, a climate of most happily tempered seasons. And as to the city, this is the way in which it was laid out at that time. In the first place, the Acropolis, as it existed then, was different from what it is now. For as it is now, the action of a single night of extraordinary rain has crumbled it away and made it bare of soil when earthquakes occurred simultaneously with the third of the disastrous floods which preceded the destructive deluge in the time of Deuealion. But in its former extent, at an earlier period, it went down towards the Eridanus and the Ilissus, and embraced within it the Pnyx, and had the Lyeabettus as its boundary over against the Pnyx; and it was all rich in soil and, save for a small space, level on the top. And its outer parts, under its slopes, were inhabited by the craftsmen and by such of the husbandmen as had their farms close by; but on the topmost part only the military class by itself had its dwellings round about the temple of Athene and Hephaestus, surrounding themselves with a single ring-fence, which formed, as it were, the enclosure of a single dwelling. On the northward side of it they had established their public dwellings and winter mess-rooms, and all the arrangements in the way of buildings which were required for the community life of themselves and the priests; but all was devoid of gold or silver, of which they made no use anywhere; on the contrary, they aimed at the mean between luxurious display and meanless, and built themselves tasteful houses, wherein they and their children's children grew old and handed them on in succession unaltered to others like themselves. As for the southward parts, when they vacated their gardens and gymnasia and mess-rooms as was natural in summer, they used them for these purposes. And near the place of the present Acropolis there was one spring – which was choked up by the earthquakes so that but small tricklings of it are now left round about but to the men of that time it afforded a plentiful stream for them all, being well tempered both for winter and summer. In this fashion, then, they dwelt, acting as guardians of their own citizens and as leaders, by their own consent, of the rest of the Greeks; and they watched carefully that their own numbers, of both men and women, who were neither too young nor too old to fight, should remain for all time as nearly as possible the same, namely, about 20 000.

So it was that these men, being themselves of the character described and always justly administering in some such fashion both their own land and Hellas, were famous throughout all Europe and Asia both for their bodily beauty and for the perfection of their moral excellence, and were of all men then living the most renowned. And now, if we have not lost recollection of what we heard when we were still children, we will frankly impart to you all, as friends, our story of the men who warred against our Athenians, what their state was and how it originally came about.

But before I begin my account, there is still a small point which I ought to explain, lest you should be surprised at frequently hearing Greek names given to barbarians. The reason of this you shall now learn. Since Solon was planning to make use of the story for his own poetry, he had found, on investigating the meaning of the names, that those Egyptians who had first written them down had translated them into their own tongue. So he himself in turn recovered the original sense of each name and, rendering it into our tongue, wrote it down so. And these very writings were in the possession of my grandfather and are actually now in mine, and when I was a child I learnt them all by heart. Therefore if the names you hear are just like our local names, do not be at all astonished, for now you know the reason for them. The story then told was a long one, and it began something like this.

Like as we previously stated concerning the allotments of the Gods, that they portioned out the whole earth, here into larger allotments and there into smaller, and provided for themselves shrines and sacrifices, even so Poseidon took for his allotment the island of Atlantis and settled therein the children whom he had begotten of a mortal woman in a region of the island of the following description. Bordering on the sea and extending through the centre of the whole island there was a plain, which is said to have been the fairest of all plains and highly fertile; and, moreover, near the plain, over against its sides. Thereon dwelt one of the natives originally sprung from the earth, Evenor by name, with his wife Leucippe; and they had for offspring an only-begotten daughter, Cleito. And when this damsel was now come to marriageable age, her mother died and also her father; and Poseidon, being smitten with desire for her, wedded her; and to make the hill whereon she dwelt impregnable he broke it off all round about; and he made circular belts of sea and land enclosing one another alternately, some greater, some smaller, two being of land and three of sea, which he carved as it were out of the midst of the island; and these belts were at even distances on all sides, so as to be impassable for man; for at that time neither ships nor sailing were as yet in existence. And Poseidon himself set in order with ease, as a god would, the central island, bringing up from beneath the earth two springs of waters, the one flowing warm from its source, the other cold, and producing out of the earth all kinds of food in plenty. And he begat five pairs of twin sons and reared them up; and when he had divided all the island of Atlantis into ten portions, he assigned to the first-born of the eldest sons his mother's dwelling and the allotment surrounding it, which was the largest and best; and him he appointed to be king over the rest, and the others to be rulers, granting, to each the rule over many men and a large tract of country. And to all of them he gave names, giving to him that was eldest and king the name after which the whole island was called and the sea spoken of as the Atlantic, because the first king who then reigned had the name of Atlas. And the name of his younger twin-brother, who had for his portion the extremity of the island near the pillars of Heracles up to the part of the country now called Gadeira after the name of that region, was Eumelus in Greek, but in the native tongue Gadeirus, – which fact may have given its title to the country. And of the pair that were born next he called the one Ampheres and the other Evaemon; and of the third pair the elder was named Mneseus and the younger Autochthon; and of the fourth pair, he called the first Elasippus and the second Mestor; and of the fifth pair, Azaes was the name given to the elder, and Diaprepês to the second. So all these, themselves and their descendants, dwelt for many generations bearing rule over many other islands throughout the sea, and holding sway besides as was previously stated, over the Mediterranean peoples as far as Egypt and Tuscany.

Now a large family of distinguished sons sprang from Atlas; but it was the eldest, who as king, always passed on the sceptre to the eldest of his sons and thus they preserved the sovereignty for many generations; and the wealth they possessed was so immense that the like had never been seen before in any royal house nor will ever easily be seen again; and they were provided with everything of which provision was needed either in the city or throughout the rest of the country. For because of their headship they had a large supply of imports from abroad, and the island itself furnished most of the requirements of daily life, – metals, to begin with, both the hard kind and the fusible kind, which are extracted by mining, and also that kind which is now known only by name but was more than a name then, there being mines of it in many places of the island, – I mean 'orichaleum', which was the most precious of the metals then known, except gold. It brought forth also in abundance all the timbers that a forest provides for the labours of carpenters; and of animals it produced a sufficiency, both of tame and wild. Moreover, it contained a very large stock of elephants; for there was an ample food-supply not only for all the other animals which haunt the marshes and lakes and rivers, or the mountains or the plains, but like wise also for this animal, which of its nature is the largest and most voracious. And in addition to all this, it produced and brought to perfection all those sweet-scented stuffs which the earth produces now, whether made of roots or herbs or trees, or of liquid gums derived from flowers or fruits. The cultivated fruit also, and the dry, which serves us for nutriment, and all the other kinds that we use for our meals – the various species of which are comprehended under the name 'vegetables', – and all the produce of trees which affords liquid and solid food and unguents, and the fruit of the orchard-trees, so hard to store, which is grown for the sake of amusement and pleasure, and all the after-dinner fruits that we serve up as welcome remedies for the sufferer from repletion, – all these that hallowed island, as it lay then beneath the sun, produced in marvellous beauty and endless abundance. And thus, receiving from the earth all these products, they furnished forth their temples and royal dwellings, their harbours and their docks, and all the rest of their country, ordering all in the fashion following.

First of all they bridged over the circles of sea which surrounded the ancient metropolis, making thereby a road towards and from the royal palace. And they had built the palace at the very beginning where the settlement was first made by their God and their ancestors; and as each king received it from his predecessor, he added to its adornment and did all he could to surpass the king before him, until finally they made of it an abode amazing to behold for the magnitude and beauty of its workmanship. For, beginning at the sea, they bored a channel right through to the outermost circle, which was three plethra in breadth, one hundred feet in depth, and fifty stades in length; and thus they made the entrance to it from the sea like that to a harbour by opening out a mouth large enough for the greatest ships to sail through. Moreover, through the circles of land, which divided those of sea, over against the bridges they opened out a channel leading from circle to circle, large enough to give passage to a single trireme; and this they roofed over above so that the sea-way was subterranean; for the lips of the land-circles were raised a sufficient height above the level of the sea. The greatest of the circles into which a boring was made for the sea was three stades in breadth. and the circle of land next to it was of equal breadth; and of the second pair of circles that of water was two stades in breadth and that of dry land equal again to the preceding one of water; and the circle which ran round the central island itself was of a stade's breadth. And this island, wherein stood the royal palace, was of five stades in diameter. Now the island and the circles and the bridge, which was a plethrum in breadth, they encompassed round about, on this side and on that, with a wall of stone; and upon the bridges on each side, over against the passages for the sea, they erected towers and gates. And the stone they quarried beneath the central island all round, and from beneath the outer and inner circles, some of it being white, some black and some red; and while quarrying it they constructed two inner docks, hollowed out and roofed over by the native rock. And of the buildings some they framed of one simple colour, in others they wove a pattern of many colours by blending the stones for the sake of ornament so as to confer upon the buildings a natural charm. And they covered with brass, as though with a plaster, all the circumference of the wall which surrounded the outermost circle; and that which encompassed the acropolis itself with orichaleum which sparkled like fire.

The royal palace within the acropolis was arranged in this manner. In the centre there stood a temple sacred to Cleito and Poseidon, which was reserved as holy ground, and encircled with a wall of gold; this being the very spot where at the beginning they had generated and brought to birth the family of the ten royal lines. Thither also they brought year by year from all the ten allotments their seasonable offerings to do sacrifice to each of those princes. And the temple of Poseidon himself was a stade in length, three plethra in breadth, and of a height which appeared symmetrical therewith; and there was something of the barbaric in its appearance. All the exterior of the temple they coated with silver, save only the pinnacles, and these they coated with gold. As to the interior, they made the roof all of ivory in appearance, variegated with gold and silver and orichaleum, and all the rest of the walls and pillars and floors they covered with orichaleum. And they placed therein golden statues, one being that of the God standing on a chariot and driving six winged steeds, his own figure so tall as to touch the ridge of the roof, and round about him a hundred Nereids on dolphins (for that was the number of them as men then believed), and it contained also many other images, the votive offerings of private men. And outside, round about the temple, there stood images in gold of all the princes, both themselves and their wives, as many as were descended from the ten kings, together with many other votive offerings both of the kings and of private persons not only from the State itself but also from all the foreign peoples over whom they ruled. And the altar, in respect of its size and its workmanship, harmonized with its surroundings; and the royal palace likewise was such as befitted the greatness of the kingdom, and equally befitted the splendour of the temples.

The springs they made use of, one kind being of

cold, another of warm water, were of abundant volume. And each kind was wonderfully well adapted for use because of the natural taste and excellence of its waters; and these they surrounded with buildings and with plantations of trees such as suited the waters; and, moreover, they set reservoirs round about, some under the open sky, and others under cover to supply hot baths in the winter; they put separate baths for the kings and for the private citizens, besides others for women, and others again for horses and all other beasts of burden, fitting out each in an appropriate manner. And the outflowing water they conducted to the sacred grove of Poseidon, which contained trees of all kinds that were of marvellous beauty and height because of the richness of the soil; and by means of channels they led the water to the outer circles over against the bridges And there they had constructed many temples for gods, and many gardens and many exercising grounds, some for men and some set apart for horses, in each of the circular belts of island, and besides the rest they had in the centre of the large island a racecourse laid out for horses, which was a stade in width, while as to length, a strip which ran round the whole circumference was reserved for equestrian contests. And round about it, on this side and on that, were barracks for the greater part of the spearmen; but the guard-house of the more trusty of them was posted in the smaller circle, which was nearer the Acropolis; while those who were the most trustworthy of all had dwellings granted to them within the acropolis round about the persons of the kings. And the shipyards were full of triremes and all the tackling that belongs to triremes, and they were all amply equipped.

Such then was the state of things round about the abode of the kings. And after crossing the three outer harbours, one found a wall which began at the sea and ran round in a circle, at a uniform distance of fifty stades from the largest circle and harbour, and its ends converged at the seaward mouth of the channel. The whole of this wall had numerous houses built on to it, while the sea-way and the largest harbour were filled with ships and merchants coming from all quarters, which by reason of their multitude caused clamour and tumult of every description and an unceasing din night and day.

Now as regards the city and the environs of the ancient dwelling we have now wellnigh completed the description as it was originally given. We must endeavour next to repeat the account of the rest of the country, what its natural character was, and in what fashion it was ordered. In the first place, then, according to the account, the whole region rose sheer out of the sea to a great height, but the part about the city was all a smooth plain, enclosing it round about, and being itself encircled by mountains which stretched as far as to the sea; and this plain had a level surface and was as a whole rectangular in shape, being 3000 stades long on either side and 2000 stades wide at its centre, reckoning upwards from the sea. And this region, all along the island, faced towards the South and was sheltered from the Northern blasts. And the mountains which surrounded it were at that time celebrated as surpassing all that now exist in number, magnitude and beauty; for they had upon them many rich villages of country folk, and streams and lakes and meadows which furnished ample nutriment to all the animals both tame and wild, and timber of various sizes and descriptions, abundantly sufficient for the needs of all and every craft.

Now as a result of natural forces, together with the labours of many kings which extended over many ages, the condition of the plain was this. It was originally a quadrangle, rectilinear for the most part, and elongated; and what it lacked of this shape they made right by means of a trench dug round about it. Now, as regards the depth of this trench and its breadth and length, it seems incredible that it should be so large as the account states, considering that it was made by hand, and in addition to all the other operations, but none the less we must report what we heard: it was dug out to the depth of a plethrum and to a uniform breadth of a stade, and since it was dug round the whole plain its consequent length was 10 000 stades.

It received the streams which came down from the mountains and after circling round the plain, and coming towards the city on this side and on that, it discharged them thereabouts into the sea. And on the inland side of the city channels were cut in straight lines, of about 100 feet in width, across the plain, and these discharged themselves into the trench on the seaward side, the distance between each being 100 stades. It was in this way that they conveyed to the city the timber from the mountains and transported also on boats the seasons' products, by cutting transverse passages from one channel to the next and also to the city. And they cropped the land twice a year, making use of the rains from heaven in the winter, and the waters that issue from the earth in summer, by conducting the streams from the trenches.

As regards their man-power, it was ordained that each allotment should furnish one man as leader of all the men in the plain who were fit to bear arms; and the size of the allotment was about ten times ten stades, and the total number of all the allotments was 60,000; and the number of the men in the mountains and in the rest of the country was countless, according to the report, and according to their districts and villages they were all assigned to these allotments under their leaders. So it was ordained that each such leader should provide for war the sixth part of a war-chariot's equipment, so as to make up 10,000 chariots in all, together with two horses and mounted men; also a pair of horses without a ear, and attached thereto a combatant with a small shield and for charioteer the rider who springs from horse to horse; and two hoplites; and archers and slingers, two of each; and light-armed slingers and javelin-men, three of each; and four sailors towards the manning of twelve hundred ships. Such then were the military dispositions of the royal City; and those of the other nine varied in various ways, which it would take a long time to tell.

Of the magistracies and posts of honour the disposition, ever since the beginning, was this. Each of the ten kings ruled over the men and most of the laws in his own particular portion and throughout his own city, punishing and putting to death whomsoever he willed. But their authority over one another and their mutual relations were governed by the precepts of Poseidon, as handed down to them by the law and by the records inscribed by the first princes on a pillar of orichaleum, which was placed within the temple of Poseidon in the centre of the island, and thither they assembled every fifth year and then alternately every sixth year – giving equal honour to both the even and the odd – and when thus assembled they took counsel about public affairs and inquired if any had in any way transgressed and gave judgement. And when they were about to give judgement they first gave pledges one to another of the following description. In the sacred precincts of Poseidon there were bulls at large, and the ten princes, being alone by themselves, after praying to the God that they might capture a victim well pleasing unto him, hunted after the bulls with staves and nooses but with no weapon of iron; and whatsoever bull they captured they led up to the pillar and cut its throat over the top of the pillar, raining down blood on the inscription. And inscribed upon the pillar, besides the laws, was an oath which invoked mighty curses upon them that disobeyed. When, then, they had done sacrifice according to their laws and were consecrating all the limbs of the bull, they mixed a bowl of wine and poured in on behalf of each one a gout of blood, and the rest they carried to the fire, when they had first purged the pillars round about. And after this they drew out from the bowl with golden ladles, and making libation over the fire swore to give judgement according to the laws upon the pillar and to punish whosoever had committed any previous transgression; and, moreover, that henceforth they would not transgress any of the writings willingly, nor govern nor submit to any governor's edict save in accordance with their father's laws. And when each of them had made this invocation both for himself and for his seed after him, he drank of the cup and offered it up as a gift in the temple of the God; and after spending the interval in supping and necessary business, when darkness came on and the sacrificial fire had died down, all the princes robed themselves in most beautiful sable vestments, and sate on the ground beside the cinders of the sacramental victims throughout the night, extinguishing all the fire that was round about the sanctuary; and there they gave and received judgement, if any of them accused any of committing any transgression. And when they had given judgement, they wrote the judgements, when it was light, upon a golden tablet, and dedicated them together with their robes as memorials. And there were many other special laws concerning the peculiar rights of the several princes, whereof the most important were these: that they should never take up arms against one another, and that, should anyone attempt to overthrow in any city their royal house, they should all lend aid, taking counsel in common, like their forerunners, concerning their policy in war and other matters, while conceding the leadership to the royal branch of Atlas; and that the king had no authority to put to death any of his brother-princes save with the consent of more than half of the ten.

Such was the magnitude and character of the power which existed in those regions at that time; and this power the God set in array and brought against these regions of ours on some such pretext as the following, according to the story. For many generations, so long as the inherited nature of the God remained strong in them, they were submissive to the laws and kindly disposed to their divine kindred. For the intents of their hearts were true and in all ways noble, and they showed gentleness joined with wisdom in dealing with the changes and chances of life and in their dealings one with another. Consequently they thought scorn of everything save virtue and lightly esteemed their rich possessions, bearing with ease the burden, as it were, of the vast volume of their gold and other goods; and thus their wealth did not make them drunk with pride so that they lost control of themselves and went to ruin; rather, in their soberness of mind they clearly saw that all these good things are increased by general amity combined with virtue, whereas the eager pursuit and worship of these goods not only causes the goods themselves to diminish but makes virtue also to perish with them As a result, then, of such reasoning and of the continuance of their divine nature all their wealth had grown to such a greatness as we previously described. But when the portion of divinity within them was now becoming faint and weak through being ofttimes blended with a large measure of mortality, whereas the human temper was becoming dominant, then at length they lost their comeliness, through being unable to bear the burden of their possessions, and beeame ugly to look upon, in the eyes of him who has the gift of sight; for they had lost the fairest of their goods from the most precious of their parts; but in the eyes of those who have no gift of perceiving what is the truly happy life, it was then above all that they appeared to be superlatively fair and blessed, filled as they were with lawless ambition and power. And Zeus, the God of gods, who reigns by law, inasmuch as he has the gift of perceiving such things, marked how this righteous race was in evil plight, and desired to inflict punishment upon them, to the end that when chastised they might strike a truer note. Wherefore he assembled together all the gods into that abode which they honour most, standing as it does at the centre of all the Universe, and beholding all things that partake of generation; and when he had assembled them he spake thus: . . .

TIMAEUS

CRIT. Listen then, Socrates, to a tale which, though passing strange, is yet wholly true, as Solon, the wisest of the Seven, once upon a time declared. Now Solon – as indeed he often says himself in his poems – was a relative and very dear friend of our great-grandfather Dropides; and Dropides told our grandfather Critias – as the old man himself, in turn, related to us – that the exploits of this city in olden days, the record of which had perished through time and the destruction of its inhabitants, were great and marvellous, the greatest of all being one which it would be proper for us now to relate both as a payment of our debt of thanks to you and also as a tribute of praise, chanted as it were duly and truly, in honour of the Goddess on this her day of Festival.

SOC. Excellent! But come now, what was this exploit described by Critias, following Solon's report, as a thing not verbally recorded, although actually performed by this city long ago?

CRIT. I will tell you: it is an old tale, and I heard it from a man not young. For indeed at that time, as he said himself. Critias was already close upon ninety years of age, while I was somewhere about ten; and it chanced to be that day of the Apaturia which is called 'Cureotis.' The ceremony for boys which was always customary at the feast was held also on that occasion, our fathers arranging contests in recitation. So while many poems of many poets were declaimed, since the poems of Solon were at that time new, many of us children chanted them. And one of our fellow-tribesmen – whether he really thought so at the time or whether he was paying a compliment to Critias – declared that in his opinion Solon was not only the wisest of men in all else, but in poetry also he was of all poets the noblest. Whereat the old man (I remember the scene well) was highly pleased and said with a smile, 'If only, Amynander, he had not taken up poetry as a by-play but had worked hard at it like others, and if he had completed the story he brought here from Egypt, instead of being forced to lay it aside owing to the seditions and all the other evils he found here on his return, – Why then, I say, neither Hesiod nor Homer nor any other poet would ever have proved more famous than he.' 'And what was the story, Critias?' said the other. 'Its subject,' replied Critias, 'was a very great exploit, worthy indeed to be accounted the most notable of all exploits, which was performed by this city, although the record of it has not endured until now owing to lapse of time and the destruction of those who wrought it.' 'Tell us from the beginning,' said Amynander, 'what Solon related and how, and who were the informants who vouched for its truth.'

In the Delta of Egypt,' said Critias, 'where, at its head, the stream of the Nile parts in two, there is a certain district called the Saitie. The chief city in this district is Sais – the home of King Amasis, – the founder of which, they says is a goddess whose Egyptian name is Neith, and in Greek, as they assert, Athena. These people profess to be great lovers of Athens and in a measure akin to our people here. And Solon said that when he travelled there he was held in great esteem amongst them; moreover, when he was questioning such of their priests as were most versed in ancient lore about their early history, he discovered that neither he himself nor any other Greek knew anything at all, one might say, about such matters. And on one occasion, when he wished to draw them on to discourse on ancient history, he attempted to tell them the most ancient of our traditions, concerning Phoroneus, who was said to be the first man, and Niobe; and he went on to tell the legend about Deucalion and Pyrrha after the Flood, and how they survived it, and to give the genealogy of their descendants; and by recounting the number of years occupied by the events mentioned he tried to calculate the periods of time. Whereupon one of the priests, a prodigiously old man, said, 'O Solon, Solon, you Greeks are always children: there is not such a thing as an old Greek.' And on hearing this he asked, 'What mean you by this saying?' And the priest replied, 'You are young in soul, every one of you. For therein you possess not a single belief that is ancient and derived from old tradition, nor yet one science that is hoary with age. And this is the cause thereof: There have been and there will be many and divers destructions of mankind, of which the greatest are by fire and water, and lesser ones by countless other means. For in truth the story that is told in your country as well as ours, how once upon a time Phaethon, son of Helios, yoked his father's chariot, and, because he was unable to drive it along the course taken by his father, burnt up all that was upon the earth and himself perished by a thunderbolt, – that story, as it is told, has the fashion of a legend, but the truth of it lies in the occurrence of a shifting of the bodies in the heavens which move round the earth, and a destruction of the things on the earth by fierce fire, which recurs at long intervals. At such times all they that dwell on the mountains and in high and dry places suffer destruction more than those who dwell near to rivers or the sea; and in our ease the Nile, our Saviour in other ways, saves us also at such times from this calamity by rising high. And when, on the other hand, the Gods purge the earth with a flood of waters, all the herdsmen and shepherds that are in the mountains are saved, but those in the cities of your land are swept into the sea by the streams; whereas in our country neither then nor at any other time does the water pour down over our fields from above, on the contrary it all tends naturally to well up from below. Hence it is, for these reasons, that what is here preserved is reckoned to be most ancient; the truth being that in every place where there is no excessive heat or cold to prevent it there always exists some human stock, now more, now less in number. And if any event has occurred that is noble or great or in any way conspicuous, whether it be in your country or in ours or in some other place of which we know by report, all such events are recorded from of old and preserved here in our temples; whereas your people and the others are but newly equipped every time, with letters and all such arts as civilized States require; and when, after the usual interval of years, like a plague, the flood from heaven comes sweeping down afresh upon your people, it leaves none of you but the unlettered and uncultured, so that you become young as ever, with no knowledge of all that happened in old times in this land or in your own. Certainly the genealogies which you related just now, Solon, concerning the people of your country, are little better than children's tales; for, in the first place, you remember but one deluge, though many had occurred previously; and next, you are ignorant of the fact that the noblest and most perfect race amongst men were born in the land where you now dwell, and from them both you yourself are sprung and the whole of your existing city, out of some little seed that chanced to be left over; but this has escaped your notice because for many generations the survivors died with no power to express themselves in writing. For verily at one time, Solon, before the greatest destruction by water, what is now the Athenian State was the bravest in war and supremely well organized also in all other respects It is said that it possessed the most splendid works of art and the noblest polity of any nation under heaven of which we have heard tell.'

Upon hearing this, Solon said that he marvelled, and with the utmost eagerness requested the priest to recount for him in order and exactly all the facts about those citizens of old. The priest then said: 'I begrudge you not the story, Solon; nay, I will tell it, both for your own sake and that of your city, and most of all for the sake of the Goddess who has adopted for her own both your land and this of ours, and has nurtured and trained them, – yours first by the space of a thousand years, when she had received the seed of you from Gê and Hephaestus, and after that ours. And the duration of our civilization as set down in our sacred writings is 8000 years. Of the citizens, then, who lived 9000 years ago, I will declare to you briefly certain of their laws and the noblest of the deeds they performed: the full account in precise order and detail we shall go through later at our leisure, taking the actual writings. To get a view of their laws, look at the laws here; for you will find existing here at the present time many examples of the laws which then existed in your city. You see, first, how the priestly class is separated off from the rest; next, the class of craftsmen, of which each sort works by itself without mixing with any other; then the classes of shepherds, hunters, and farmers, each distinct and separate. Moreover, the military class here, as no doubt you have noticed, is kept apart from all the other classes, being enjoined by the law to devote itself solely to the work of training for war. A further feature is the character of their equipment with shields and spears; for we were the first of the peoples of Asia to adopt these weapons, it being the Goddess who instructed us, even as she instructed you first of all the dwellers in yonder lands. Again, with regard to wisdom, you perceive, no doubt, the law here, – how much attention it has devoted from the very beginning to the Cosmic Order, by discovering all the effects which the divine causes produce upon human life, down to divination and the art of medicine which aims at health, and by its mastery also of all the other subsidiary studies. So when, at that time, the Goddess had furnished you, before all others, with all this orderly and regular system, she established your State, choosing the spot wherein you were born since she perceived therein a climate duly blended, and how that it would bring forth men of supreme wisdom. So it was that the Goddess, being herself both a lover of war and a lover of wisdom, chose the spot which was likely to bring forth men most like unto herself, and this first she established. Wherefore you lived under the rule of such laws as these, – yea, and laws still better, – and you surpassed all men in every virtue, as became those who were the offspring and nurslings of gods. Many, in truth, and great are the achievements of your State, which are a marvel to men as they are here recorded; but there is one which stands out above all both for magnitude and for nobleness. For it is related in our records how once upon a time your State stayed the course of a mighty host, which, starting from a distant point in the Atlantic ocean, was insolently advancing to attack the whole of Europe, and Asia to boot. For the ocean there was at that time navigable; for in front of the mouth which you Greeks call, as you say, 'the pillars of Heracles,' there lay an island which was larger than Libya and Asia together; and it was possible for the travellers of that time to cross from it to the other islands, and from the islands to the whole of the continent over against them which encompasses that veritable ocean. For all that we have here, lying within the mouth of which we speak, is evidently a haven having a narrow entrance; but that yonder is a real ocean, and the land surrounding it may most rightly be called, in the fullest and truest sense, a continent. Now in this island of Atlantis there existed a confederation of kings, of great and marvellous power, which held sway over all the island, and over many other islands also and parts of the continent; and, moreover, of the lands here within the Straits they ruled over Libya as far as Egypt, and over Europe as far as Tuscany. So this host, being all gathered together, made an attempt one time to enslave by one single onslaught both your country and ours and the whole of the territory within the Straits. And then it was, Solon, that the manhood of your State showed itself conspicuous for valour and might in the sight of all the world. For it stood pre-eminent above all in gallantry and all warlike arts, and acting partly as leader of the Greeks, and partly standing alone by itself when deserted by all others, after encountering the deadliest perils, it defeated the invaders and reared a trophy; whereby it saved from slavery such as were not as yet enslaved, and all the rest of us who dwell within the bounds of Heracles it ungrudgingly set free. But at a later time there occurred portentous earthquakes and floods, and one grievous day and night befell them, when the whole body of your warriors was swallowed up by the earth, and the island of Atlantis in like manner was swallowed up by the sea and vanished; wherefore also the ocean at that spot has now become impassable and unsearchable, being blocked up by the shoal mud which the island created as it settled down.'

You have now heard, Socrates, in brief outline, the account given by the elder Critias of what he heard from Solon; and when you were speaking yesterday about the State and the citizens you were describing, I marvelled as I called to mind the facts I am now relating, reflecting what a strange piece of fortune it was that your description coincided so exactly for the most part with Solon's account. I was loth, however, to speak on the instant; for owing to lapse of time my recollection of his account was not sufficiently clear. So I decided that I ought not to relate it until I had first gone over it all carefully in my own mind.

Consequently, I readily consented to the theme you proposed yesterday, since I thought that we should be reasonably well provided for the task of furnishing a satisfactory discourse – which in all such cases is the greatest task. So it was that, as Hermocrates has said, the moment I left your place yesterday I began to relate to them the story as I recollected it, and after I parted form them I pondered it over during the night and recovered, as I may say, the whole story. Marvellous, indeed, is the way in which the lessons of one's childhood 'grip the mind', as the saying is. For myself, I know not whether I could recall to mind all that I heard yesterday; but as to the account I heard such a great time ago, I should be immensely surprised if a single detail of it has escaped me. I had then the greatest pleasure and amusement in hearing it, and the old man was eager to tell me, since I kept questioning him repeatedly, so that the story is stamped firmly on my mind like the encaustic designs of an indelible painting. Moreover, immediately after daybreak I related this same story to our friends here, so that they might share in my rich provision of discourse.

Now, therefore, – and this is the purpose of all that I have been saying, – I am ready to tell my tale, not in summary outline only but in full detail just as I heard it. And the city with its citizens which you described to us yesterday, as it were in a fable, we will now transport hither into the realm of fact; for we will assume that the city is that ancient city of ours, and declare that the citizens you conceived are in truth those actual progenitors of ours, of whom the priest told. In all ways they will correspond, nor shall we be out of tune if we affirm that those citizens of yours are the very men who lived in that age. Thus, with united effort, each taking his part, we will endeavour to the best of our powers to do justice to the theme you have prescribed. Wherefore, Socrates, we must consider whether this story is to our mind, or we have still to look for some other to take its place.

Soc. What story should we adopt, Critias, in preference to this? For this story will be admirably suited to the festival of the Goddess which is now being held, because of its connexion with her; and the fact that it is no invented fable but genuine history is all-important. How, indeed, and where shall we discover other stories if we let these slip? Nay, it is impossible. You, therefore, must now deliver your discourse (and may Good Fortune attend you!), while I, in requital for my speech of yesterday, must now keep silence in my turn and hearken.

From *Plato Timaeus, Critias*
The Loeb Classical Library
Plato VII
Translated by The Rev. H.G. Bury, Litt. D.
Harvard University Press 1952

Appendix 2

List of fossils

Foraminifera from Archangelos Vouno and from the locality Akrotiri I

Benthic foraminifera

	Frequency (per cent)
Ammonia beccarii s.s. (Linné, 1758 = *Nautilus beccarii*)	1.0
Ammonia beccarii, var. *sobrinus* (Shupack, 1934 = *Rotalia beccarii*, var. *sobrina*)	17.7
Anomalinoides ornatus (Costa, 1850 = *Nonionina ornata*)	0.8
Asterigerinata mamilla (Williamson, 1858 = *Rotalia mamilla*)	4.6
Asterigerinata planorbis (d'Orbigny, 1846 = *Asterigina planorbis*)	0.1
Biloculinella depressa (d'Orbigny, 1826 = *Biloculina depressa*)	0.1
Bulimina gibba Fornasini, 1902	0.2
Cancris auriculus (Fishtel & Moll, 1798 = *Nautilus auricula*)	0.1
Cassidulina laevigata d'Orbigny, 1826	1.8
Cassidulina obtusa Williamson, 1858	0.5
Cibicides lobatulus (Walker & Jacob, 1798 = *Nautilus lobatulus*)	11.9
Cibicides refulgens Montfort, 1808	0.5
Discorbinella sp.	0.1
Elphidium aculeatum (d'Orbigny, 1846 = *Polystomella aculeata*)	0.7
Elphidium advenum (Cushman, 1922 = *Polystomella advena*)	0.4
Elphidium articulatum (d'Orbigny, 1839 = *Polystomella articulatum*)	0.2
Elphidium complanatum (d'Orbigny, 1839 = *Polystomella complanata*)	1.2
Elphidium complanatum, var. *tyrrhenianum* Accordi, 1951	0.6
Elphidium crispum (Linné, 1758 = *Nautilus crispum*)	5.2
Elphidium fichtellianum (d'Orbigny, 1846 = *Polystomella fichtelliana*)	2.2
Elphidium macellum (Fishtel & Moll, 1798 = *Nautilus macellus*)	4.6
Elphidium sp.	0.1
Eponides turgidus Phleger & Parker, 1951	0.1
Fissurina sp.	0.1
Gavelinopsis praegeri (Heron-Allen & Earland, 1913 = *Discorbina praegeri*)	0.1

Globocassidulina subglobosa (Brady, 1881 = *Cassidulina subglobosa*)	1.9
Globulina gibba d'Orbigny, 1826	0.5
Guttulina sp.	0.5
Gyroidina soldanii d'Orbigny, 1826	0.2
Heterolepa pseudoungeriana (Cushman, 1922 = *Truncatulina pseudoungeriana*)	2.9
Heterolepa ungeriana (d'Orbigny, 1846 = *Rotalia ungeriana*)	0.4
Laryngosigma lactea (Walker & Jacob, 1798 = *Serpula lactea*)	0.5
Melonis barleeanus (Williamson, 1858 = *Nonionina barleeana*)	4.3
Miliolinella cf. *M. fichteliana* (d'Orbigny, 1839 = *Triloculina fichteliana*)	0.1
Miliolinella subrotunda (Montagu, 1803 = *Vermiculum subrotundum*)	2.5
Neoconcorbina millettii (Wright, 1910 = *Discorbis millettii*)	6.4
Ninion limbum (d'Orbigny, 1826 = *Nonionina limba*)	2.4
Osangularia culter (Parker & Jones, 1865 = *Planorbulina culter*)	1.0
Paromalina bilateralis Loeblich & Tappan, 1957	0.1
Patellina corrugata Williamson, 1858	0.1
Planorbulina mediterranensis d'Orbigny, 1826	0.2
Planulina ariminensis d'Orbigny, 1826	1.1
Pyrgo tubulosa (Costa, 1856 = *Biloculina tubulosa*)	0.2
Pyrgo elongata (d'Orbigny, 1826 = *Biloculina elongata*)	0.2
Quinqueloculina lamarchiana d'Orbigny, 1839	0.2
Quinqueloculina lata Terquem, 1876	0.2
Quinqueloculina longirostra d'Orbigny, 1826	0.1
Quinqueloculina oblonga (Montagu, 1803 = *Vermiculum oblongum*)	0.7
Quingueloculina padana Perconig, 1954	0.1
Quinqueloculina seminulum s.l. (Linné, 1758 = *Serpula seminulum*)	6.3
Quinqueloculina venusta Karrer, 1868	0.1
Quinqueloculina vulgaris d'Orbigny, 1826	1.2
Rosalina bradyi (Cushman, 1915 = *Discorbis globularis*, var. *bradyi*)	0.1
Rosalina cf. *R. bradyi* (Cushman, 1915 = *Discorbis globularis*, var. *bradyi*)	0.7
Rosalina carnivora Todd, 1965	0.1
Rosalina concinna d'Orbigny, 1826	0.2
Rosalina globularis d'Orbigny, 1826	1.0
Rosalina aff. *R. macropora* (Hofker, 1951 = *Discopulvinolina macropora*)	2.3
Sigmomorphina semitecta (Reuss, 1867 = *Polymorphina semitecta*)	0.1
Spiroloculina depressa d'Orbigny, 1826	0.1
Spiroloculina excavata d'Orbigny, 1846	0.4
Stomatorbina concentrica (Parker & Jones, 1864 = *Pulvinulina concentrica*)	1.3
Textularia candeiana d'Orbigny, 1839	0.1
Textularia pseudogramen Chapman & Parr, 1937	0.4
Textularia pseudorugosa Lacroix, 1931	0.1

Trifarina angulosa (Williamson, 1858 = *Uvigerina angulosa*) 0.5
Triloculina inflata d'Orbigny, 1826 0.8
Triloculina oblonga (Montagu, 1803 = *Vermiculum oblongum*) 0.4
Triloculina trigonula (Lamarck, 1804 = *Miliolites trigonula*) 0.1
Uvigerina flintii Cushman, 1923 0.2
Valvulineria complanata (d'Orbigny, 1846 = *Rosalina complanata*) 0.1
Andere Miliolidae 0.5
Indeterminate 1.7
Material: 831 benthic specimens.

Planktonic foraminifera

Orbulina universa d'Orbigny, 1839 45.0
Neogloboquadrina pachyderma
(Ehrenberg, 1861 = *Aristerospira pachyderma*) 22.5
Globigerinoides conglubatus (Brady, 1879 = *Globigerina conglubata*) } 20.0
Globigerinoides ruber (d'Orbigny, 1839 = *Globigerina ruber*) }
Globigerinella siphonifera (d'Orbigny, 1839 = *Globigerina siphonifera*) 7.5
Globorotalia inflata (d'Orbigny, 1839 = *Globigerina inflata*) 5.0
Material: 40 planktonic specimens.

Foraminifera from the locality Akrotiri II

Benthic Foraminifera

Ammonia beccarii s.l. (Linné, 1758 = *Nautilus beccarii*) 2.7
Asterigerinata planorbis (d'Orbigny, 1846 = *Asterigina planorbis*) 2.7
Cassidulina laevigata d'Orbigny, 1826 2.7
Cibicides lobatulus (Walker & Jacob, 1798 = *Nautilus lobatulus*) 2.7
Nummoloculina contraria (d'Orbigny, 1846 = *Biloculina contraria*) 2.7
Pyrgo comata (Brady, 1881 = *Biloculina comata*) 2.7
Pyrgo elongata (d'Orbigny, 1826 = *Biloculina elongata*) 5.4
Pyrgo tubulosa (Costa, 1856 = *Biloculina tubulosa*) 2.7
Quinqueloculina lamarchina d'Orbigny, 1839 2.7
Quinqueloculina lata Terquem, 1876 5.4
Quinqueloculina padana Perconig, 1954 14.5
Quinqueloculina seminulum (Linné, 1758 = *Serpula seminulum*) 32.4
Quinqueloculina vulgaris d'Orbigny, 1846 10.8
Sphaeroidina bulloides d'Orbigny, 1826 5.4
Triloculina inflata d'Orbigny, 1826 5.4
Material: 37 benthic specimens.

Planktonic foraminifera

Species	%
Globorotalia inflata (d'Orbigny, 1839 = *Globigerina inflata*)	8.0
Globigerinella siphonifera (d'Orbigny, 1839 = *Globigerina siphonifera*)	4.0
Globigerinoides conglubatus (Brady, 1879 = *Globigerina conglubata*)	4.0
Orbulina universa d'Orbigny, 1839	68.0
Indeterminate	16.0

Material: 25 planktonic specimens.

Foraminifera from Cape Lumaravi

Benthic foraminifera

Species	%
Ammonia beccarii, var. *sobrinus* (Shupack, 1934 = *Rotalia beccarii*, var. *sobrina*)	0.4
Anomalinoides ornatus (Costa, 1850 = *Nonionina ornata*)	14.1
Bolivina pseudoplicata Heron-Allen & Earland, 1930	0.2
Bulimina aculeata d'Orbigny, 1826	0.2
Cassidulina crassa d'Orbigny, 1839	0.4
Cassidulina laevigata d'Orbigny, 1826	0.2
Chilostomella mediterranensis Cushman & Todd, 1949	4.5
Cibicides fletcheri Galloway & Wissler, 1927	0.2
Cibicides lobatulus (Walker & Jacob, 1798 = *Nautilus lobatulus*)	0.9
Cibicidoides pachydermus (Rzehak, 1953 = *Truncatulina pachyderma*)	12.2
? *Discorbis* sp.	1.4
Eggerella bradyi (Cushman, 1911 = *Verneilina bradyi*)	0.4
Elphidium complanatum (d'Orbigny, 1839 = *Polystomella complanata*)	1.3
Elphidium complanatum, var. *tyrrhenianum* Accordi, 1951	0.2
Elphidium macellum (Fishtel & Moll, 1798 = *Nautilus macellus*)	0.2
Epistominella sp.	0.2
Fissurina orbignyana Seguenza, 1826	0.2
Florilus asterizans (Fichtel & Moll, 1798 = *Nautilus asterizans*)	0.4
Globocassidulina subglobosa (Bradyi, 1881 = *Cassidulina subglobosa*)	1.7
Gyroidina neosoldanii Brotzen, 1936	0.9
Gyroidina soldanii d'Orbigny, 1826	12.7
Heterolepa pseudoungeriana (Cushman, 1922 = *Truncatulina pseudoungeriana*)	0.7
Hyalinea balthica (Schroeter, 1783 = *Nautilus balthicus*)	0.2
Kareriella bardyi (Cushman, 1911 = *Gaudryina bradyi*)	0.2
Lenticulina orbicularis (d'Orbigny, 1826 = *Robulina orbicularis*)	0.2
Lenticulina thalmanni (Hessland, 1943 = *Robulus thalmanni*)	0.2
Loxostomoides sp.	0.2
Martinotinella sp.	0.2
Melonis barleeanus (Williamson, 1858 = *Nonionina barleeana*)	17.5

Nonion limbum (d'Orbigny, 1826 = *Nonionina limba*) 2.6
Oridorsalis stellatus (Silvestri, 1898 = *Truncatulina tenera* ?, var. *stellata*) 0.4
Pyrgo elongata (d'Orbigny, 1826 = *Biloculina elongata*) 0.2
Rosalina cf. *R. bradyi* (Cushman, 1915 = *Discorbis globularis*, var. *bradyi*) 0.2
Sphaeroidina bulloides d'Orbigny, 1826 1.3
Stomatorbina concentrica
(Parker & Jones, 1864 = *Pulvinulina concentrica*) 0.2
Trifarina angulosa (Williamson, 1858 = *Uvigerina angulosa*) 0.4
Uvigerina flintii Cushman, 1923 1.1
Uvigerina mediterranea Hofker, 1932 21.5
Uvigerina peregrina Cushman, 1923 0.7
Uvigerina proscidea Schwager, 1866 0.2
Valvulineria complanata (d'Orbigny, 1846 = *Rosalina complanata*) 0.6
Indeterminate 0.7
Material: 543 benthic specimens.

Planktonic foraminifera

Globigerina bulloides d'Orbigny, 1826 1.0
Globigerinoides conglubatus (Brady, 1879 = *Globigerina conglubata*)
Globigerinoides ruber (d'Orbigny, 1839 = *Globigerina ruber*)
Globorotalia crassaformis
(Galloway & Wissler, 1927 = *Globigerina crassaformis*) } 0.1
Globorotalia inflata (d'Orbigny, 1839 = *Globigerina inflata*) 24.4
Globorotalia scitula (Brady, 1882 = *Pulvinulina scitula*) 0.2
Neogloboquadrina dutertrei (d'Orbigny, 1839 = *Globigerina dutertrei*) 0.3
Neogloboquadrina pachyderma
(Ehrenberg, 1861 = *Aristerospira pachyderma*) 7.4
Orbulina universa d'Orbigny, 1839 34.8
Sphaeroidinella dehiscens
(Parker & Jones, 1865 = *Sphaeroidina dehiscens*) 0.1
Turborotalita quinqueloba (Natland, 1938 = *Globigerina quinqueloba*) 0.9
Indeterminate 1.4
Material: 1203 planktonic specimens.

Appendix 3

List of flora

Compiled after Thomas Raus in: Helmut Schmalfuss: *Santorin: Leben in Schutt und Asche* (1991).

A. Pteridophyta
(Spore-bearing plants)

Equisetaceae
(Horsetail family)
Equisetum ramosissimum

Ophioglossaceae
(Adder's-tongue family)
Ophioglossum lusitanicum

Polypodiaceae
(Staghorn fern family)
Adiantum capillus-veneris
Anogramma leptophylla
Asplenium ceterach
Asplenium obovatum
Asplenium onopteris
Cheilanthes acrostica
Cheilanthes maderensis
Cheilanthes vellea
Polypodium cambricum
Pteridium aquilinum

Selaginellaceae
(Lesser club moss family)
Selaginella denticulata

B. Gymnospermae
(Naked-seed-bearing plants)

Cupressaceae
(Cypress family)
Cupressus sempervirens
Juniperus phoenicea

Ephedraceae
(Ephedra family)
Ephedra foeminea

Pinaceae
(Pine family)
Pinus brutia

C. Dicotyledones
(Dicots)

Aizoaceae
(Carpetweed family)
Aptenia cordifolia
Carpobrotus acinaciformis
Lampranthus sp.
Mesembryanthemum crystallinum
Mesembryanthemum nodiflorum

Amaranthaceae
(Amaranth family)
Amaranthus blitoides

Amaranthus caudatus
Amaranthus deflexus
Amaranthus graecizans
Amaranthus retroflexus
Amaranthus viridis

Anacardiaceae
(Sumac family)
Pistacia lentiscus

Apocynaceae
(Dogbane family)
Nerium oleander

Asclepiadaceae
(Milkweed family)
Asclepias fruticosa

Boraginaceae
(Borage family)
Alkanna tinctoria
Anchusa aegyptiaca
Anchusa hybrida
Echium angustifolium
Echium arenarium
Echium lycopsis
Echium parviflorum
Heliotropium dolosum
Heliotropium europaeum
Heliotropium suaveolens
Heliotropium supinum
Lithospermum sibthorpianum
Myosotis arvensis
Myosotis incrassata
Myosotis ramosissima
Nonea pulla

Cactaceae
(Cactus family)
Opuntia ficus-indica

Campanulaceae
(Bellflower family)
Campanula erinus
Legousia hybrida

Cannabaceae
(Hemp family)
Cannabis sativa

Capparidaceae
(Caper family)
Capparis spinosa

Caprifoliaceae
(Honeysuckle family)
Lonicera etrusca

Caryophyllaceae
(Pink family)
Agrostemma githago
Arenaria leptoclados
Cerastium comatum
Cerastium glomeratum
Cerastium glutinosum
Cerastium semidecandrum
Herniaria cinerea
Herniaria hirsuta
Paronychia argentea
Paronychia echinulata
Paronychia macrosepala
Petrorhagia velutina
Polycarpon tetraphyllum
Sagina apetala
Sagina maritima
Silene behen
Silene colorata
Silene cretica
Silene cythnia
Silene gallica
Silene nocturna
Silene sartorii
Silene sedoides

Silene vulgaris
Spergula arvensis
Spergularia bocconei
Stellaria pallida
Vaccaria hispanica

Chenopodiaceae
(Goosefoot family)
Atriplex halimus
Atriplex portulacoides
Atriplex recurva
Beta maritima
Chenopodium murale
Chenopodium opulifolium
Salsola kali
Sarcocornia fruticosa

Cistaceae
(Rock rose family)
Cistus creticus
Cistus salviifolius
Fumana arabica
Fumana thymifolia
Helianthemum aegyptiacum
Helianthemum salicifolium
Tuberaria guttata

Compositae
(Aster family)
Aetheorhiza bulbosa
Ambrosia maritima
Andryala integrifolia
Anthemis rigida
Anthemis tomentosa
Anthemis werneri
Artemisia arborescens
Argyranthemum frutescens
Asteriscus spinosus
Atractylis cancellata
Bellium minutum
Calendula arvensis
Calendula bicolor
Carduus pycnocephalus
Carlina corymbosa
Carthamus lanatus
Carthamus leucocaulos
Centaurea mixta
Chondrilla juncea
Chrysanthemum coronarium
Chrysanthemum segetum
Cichorium spinosum
Conyza bonariensis
Crepis foetida
Crepis multiflora
Crepis sancta
Crupina crupinastrum
Diotis maritima
Diffrichia graveolens
Diffrichia viscosa
Echinops spinosissimus
Filago aegaea
Filago contracta
Filago cretensis
Filago eriocephala
Filago gallica
Filago pyramidata
Filago vulgaris
Hedypnois cretica
Helichrysum barrelieri
Helichrysum italicum
Hymenonema graecum
Hyoseris scabra
Hypochoeris achyrophorus
Hypochoeris glabra
Lactuca serriola
Leontodon tuberosus
Matricaria chamomilla
Onopordon argolicum
Onopordon caulescens
Phagnalon graecum
Picris altissima
Picris pauciflora
Picnomon acarna
Reichardia picroides

Scolymus hispanicus
Scorzonera eximia
Senecio glaucus
Senecio vulgaris
Silybum marianum
Sonchus asper
Sonchus oleraceus
Sonchus tenerrimus
Taraxacum laevigatum
Taraxacum megalorrhizon
Tolpis barbata
Tragopogon sinuatus
Urosperum picroides

Convolvulaceae
(Morning glory family)
Convolvulus althaeoides
Convolvulus arvensis
Convolvulus dorycnium
Convolvulus oleifolius
Convolvulus siculus
Cuscuta palaestina

Crassulaceae
(Stonecrop family)
Crassula alata
Sedum amplexicaule
Sedum hispanicum
Sedum litoreum
Sedum rubens
Sedum sediforme
Umbilicus horizontalis
Umbilicus rupestris

Cruciferae
(Mustard family)
Alyssum simplex
Alyssum umbellatum
Arabidopsis thaliana
Arabis verna
Biscutella didyma
Brassica rapa
Brassica tournefortii
Bunias erucago
Cakile maritima
Capsella bursa-pastoris
Cardaria draba
Clypeola jonthlaspi
Didesmus aegyptius
Erophila praecox
Eruca sativa
Erysimum senoneri
Hirschfeldia incana
Lobularia libyca
Malcolmia chia
Malcolmia flexuosa
Matthiola incana
Matthiola sinuata
Raphanus raphanistrum
Rapistrum rugosum
Sisymbrium irio
Sisymbrium orientale
Sisymbrium polyceratium
Teesdalia coronopifolia

Cucurbitaceae
(Gourd family)
Bryonia cretica

Ericaceae
(Heath family)
Erica manipuliflora

Euphorbiaceae
(Spurge family)
Chrozophora obliqua
Euphorbia acanthothamnos
Euphorbia chamaesyce
Euphorbia dendroides
Euphorbia paralias
Euphorbia peplis
Euphorbia peplus
Euphorbia terracina
Mercurialis annua

Fagaceae
(Beech family)
Quercus coccifera

Frankeniaceae
(Frankenia family)
Frankenia hirsuta

Gentianaceae
(gentian family)
Blackstonia perfoliata
Centaurium tenuiflorum

Geraniaceae
(Geranium family)
Erodium botrys
Erodium chium
Erodium ciconium
Erodium cicutarium
Erodium gruinum
Erodium laciniatum
Erodium neuradifolium
Geranium molle
Geranium robertianum
Geranium rotundifolium
Pelargonium graveolens

Guttiferae
(Garcinia family)
Hypericum triquetrifolium

Labiatae
(Mint family)
Ajuga iva
Ballota acetabulosa
Coridothymus capitatus
Lamium amplexicaule
Marrubium vulgare
Mentha longifolia
Origanum onites
Prasium majus
Rosmarinus officinalis
Salvia fruticosa
Salvia pomifera
Salvia verbenaca
Salvia graeca
Satureja juliana
Satureja nervosa
Satureja thymbra
Sideritis lanata
Teucrium brevifolium
Teucrium capitatum
Teucrium divaricatum

Leguminosae
(Pea family)
Anthyllis hermanniae
Anthyllis vulneraria
Astragalus boeticus
Astragalus hamosus
Astragalus pelecinus
Astragalus peregrinus
Astragalus sinaicus
Bituminaria bituminosa
Calicotome villosa
Ceratonia siliqua
Coronilla scorpioides
Hedysarum spinosissimum
Hippocrepis ciliata
Hymenocarpus circinnatus
Lathyrus aphaca
Lathyrus cicera
Lathyrus clymenum
Lathyrus ochrus
Lathyrus sativus
Lathyrus saxatilis
Lathyrus setifolius
Lathyrus sphaericus
Lotus cytisoides
Lotus edulis
Lotus halophilus
Lotus ornithopodioides
Lotus peregrinus
Lupinus angustifolius

Lupinus varius
Medicago arborea
Medicago coronata
Medicago disciformis
Medicago litoralis
Medicago marina
Medicago minima
Medicago monspeliaca
Medicago orbicularis
Medicago polymorpha
Medicago praecox
Medicago rugosa
Medicago truncatula
Melilotus alba
Melilotus indica
Melilotus neapolitana
Melilotus sulcata
Onobrychis aequidentata
Onobrychis caput-galli
Ononis diffusa
Ononis ornithopodioides
Ononis pubescens
Ononis reclinata
Ornithopus compressus
Ornithopus pinnatus
Scorpiurus muricatus
Trifolium arvense
Trifolium campestre
Trifolium cherleri
Trifolium dasyurum
Trifolium glomeratum
Trifolium hirtum
Trifolium infamia-ponertii
Trifolium nigrescens
Trifolium resupinatum
Trifolium scabrum
Trifolium spumosum
Trifolium stellatum
Trifolium subterraneum
Trifolium suffocatum
Trifolium tomentosum
Trifolium uniflorum
Trigonella balansae
Trigonella coerulescens
Trigonella spruneriana
Vicia articulata
Vicia cretica
Vicia cuspidata
Vicia hybrida
Vicia lathyroides
Vicia lutea
Vicia peregrina
Vicia sativa
Vicia villosa

Linaceae
(Flax family)
Linum bienne
Linum strictum

Malvaceae
(Mallow family)
Lavatera arborea
Lavatera cretica
Malva parviflora
Malva sylvestris

Moraceae
(Mulberry family)
Ficus carica

Orobanchaceae
(Broomrape family)
Orobanche minor
Orobanche pubescens
Orobanche ramosa

Oxalidaceae
(Wood-sorrel family)
Oxalis corniculata
Oxalis pes-caprae

Papaveraceae
(Poppy family)
Fumaria bastardii
Fumaria densiflora
Fumaria judaica
Fumaria parviflora
Fumaria petteri
Glaucium flavum
Hypecoum procumbens
Papaver argemone
Papaver dubium
Papaver hybridum
Papaver rhoeas
Papaver somniferum
Roemena hybrida

Plantaginaceae
(Plantain family)
Plantago afra
Plantago albicans
Plantago arenaria
Plantago bellardii
Plantago lagopus
Plantago weldenii

Plumbaginaceae
(Leadwort family)
Limonium graecum
Limonium narbonense
Limonium sinuatum
Limonium virgatum

Polygonaceae
(Buckwheat family)
Emex spinosa
Polygonum arenastrum
Polygonum convolvulus
Polygonum equisetiforme
Polygonum maritimum
Rumex acetosella
Rumex bucephalophorus
Rumex tuberosus

Portulacaceae
(Purslane family)
Portulaca oleracea

Primulaceae
(Primrose family)
Anagallis arvensis
Asterolinon linum-stellatum

Ranunculaceae
(Buttercup family)
Anemone pavonina
Consolida ajacis
Nigella degenii
Nigella doeffleri
Ranunculus creticus
Ranunculus neapolitanus
Ranunculus paludosus

Resedaceae
(Reseda family)
Reseda alba
Reseda lutea
Reseda luteola

Rhamnaceae
(Buckthorn family)
Rhamnus lycioides

Rosaceae
(Rose family)
Amygdalus communis
Sarcopoterium spinosum

Rubiaceae
(Madder family)
Crucianella latifolia
Galium aparine
Galium murale
Galium recurvum
Galium tricornutum
Galium spurium

Galium verrucosum
Rubia tinctorum
Sheradia arvensis
Valantia hispida
Valantia muralis

Rutaceae
(Rue family)
Ruta chalepensis

Santalaceae
(Sandalwood family)
Thesium humile

Scrophulariaceae
(Figwort family)
Kickxia elatine
Linaria chalepensis
Linaria parviflora
Linaria pelisseriana
Linaria simplex
Misopates orontium
Parentucellia latifolia
Scrophularia heterophylla
Scrophularia lucida
Verbascum sinuatum
Veronica cymbalaria
Veronica hederifolia
Veronica praecox

Solanaceae
(Nightshade family)
Datura innoxia
Hyoscyamus albus
Lycium schweinfurthii
Mandragora autumnalis
Nicotiana glauca
Slanum luteum
Solanum nigrum

Tamaricaceae
(Tamarisk family)
Tamarix parviflora

Theligonaceae
(Theligonum family)
Theligonum cynocrambe

Thymelaeaceae
(Mezereum family)
Thymelaea hirsuta

Umbelliferae
(Parsley family)
Bifora testiculata
Bupleurum semicompositum
Bupleurum trichopodum
Githmum maritimum
Daucus carota
Daucus guttatus
Daucus involucratus
Eryngium maritimum
Ferula communis
Foeniculum vulgare
Lagoecia cuminoides
Pimpinella pretenderis
Pseudorlaya pumila
Scaligeria napiformis
Scandix pecten-veneris
Smyrnium olusatrum
Thapsia garganica
Tordylium apulum
Torilis leptophylla
Torilis nodosa

Urticaceae
(Nettle family)
Partietaria cretica
Partietaria judaica
Urtica pilulifera
Urtica urens

Valerianaceae
(Valerian family)
Centranthus ruber
Valerianella discoidea

Verbenaceae
(Verbena family)
Vitex agnus-castus

Zygophyllaceae
(Zygophyllum family)
Tribulus terrestris

D. Monocotyledones
(Monocots)

Amaryllidaceae
(Amaryllis family)
Pancratium maritimum

Araceae
(Arum family)
Arisarum vulgare

Cyperaceae
(Sedge family)
Cyperus rotundus

Gramineae
(Grass family)
Aegilops biuncialis
Aegilops neglecta
Aira cupaniana
Aira elegantissima
Alopecurus myosuroides
Arundo donax
Avellinia michelii
Avena barbata
Avena sterilis
Briza maxima
Bromus fasciculatus
Bromus hordeaceus
Bromus intermedi
Bromus madritensis
Bromus rigidus
Bromus rubens
Bromus tectorum
Catapodium marinum
Catapodium rigidum
Corynephorus divaricatus
Cynodon dactylon
Cynosurus echinatus
Dactylis glomerata
Elymus farctus
Festuca arundinacea
Gastridium phleoides
Haynaldia villosa
Holcus setiglumis
Hordeum bulbosum
Hordeum leporinum
Hordeum vulgare
Hyparrhenia hirta
Lagurus ovatus
Lamarckia aurea
Lolium perenne
Lolium rigidum
Lolium subulatum
Lolium temulentum
Melica minuta
Parapholis marginata
Parapholis incurva
Phalaris canariensis
Phleum exaratum
Phleum subulatum
Piptatherum miliaceum
Poa pelasgis
Polypogon subspathaceus
Psilurus incurvus
Rostraria cristata
Schismus arabicus
Setaria adhaerens
Stipa capensis
Trachynia distachya
Triplachne nitens

Vulpia ciliata
Vulpia fasciculata
Vulpia muralis

Iridaceae
(Iris family)
Crocus laevigatus
Gynandriris sisyrhinchium
Iris florentina
Iris germanica
Romulea bulbocodium

Juncaceae
(Rush family)
Juncus heldreichianus

Liliaceae
(Lily family)
Allium ampeloprasum
Allium bourgeaui
Allium cupanii
Allium guttatum
Allium neapolitanum
Allium staticiforme
Allium subhirsutum
Aloe vera
Asparagus aphyllus
Asparagus stipularis
Asphodelus aestivus
Asphodelus fistulosus
Colchicum cupanii
Gagea graeca
Muscari commutatum
Muscari comosum
Muscari cycladicum
Muscari weissii
Scilla autumnalis
Urginea maritima

Orchidaceae
(Orchid family)
Anacamptis pyramidalis
Ophrys fusca
Ophrys iricolor
Ophrys luteaOphrys scolopax
Orchis anatolica
Orchis papilionacea
Orchis sancta
Serapias vomeracea

Potamogetonaceae
(Pondwood family)
Posidonia oceanica
Ruppia cirrhosa

Zanichelliaceae
(Horned pondweed family)
Cymodocea nodosa

References

Åberg, N. (1933). *Bronzezeitliche und Früheisenzeitliche Chronologie. Teil IV. Griechenland.* Stockholm (Verlag der Akademie).

Ammianus Marcellinus (1874). *Rerum gestarum libri qui supersunt.* Leipzig (Gardthausen).

Apollonius Rhodius (1905). *Argonautica.* Leipzig (Merkel).

Apollonius Rhodius (1997). *The Argonautica* (trsl. by Peter Green). University of California Press, 474 pp.

Arvanitides, N.; Boström, K.; Kalogeropoulos, S.; Paritsis, S.; Galanopoulos, V.; Papavassiliou, C. (1990). Geochemistry of lavas, pumice and veins in drill core GPK-1, Palaea Kameni, Santorini. In: Hardy, D. A. (Ed.) *Thera and the Aegean World III.* vol 2. London (The Thera Foundation). 266–279.

Aston, M. A.; Hardy, P. G. (1990). The Pre-Minoan landscape of Thera: a preliminary statement. In: Hardy, D. A. (Ed.) *Thera and the Aegean World III.* vol. 2. London (The Thera Foundation). 348–361.

Baillie, M. G. L. (1990). Irish tree rings and an event in 1628 BC. In: Hardy, D. A. (Ed.) *Thera and the Aegean World III.* vol. 3. London (The Thera Foundation). 160–166.

Baillie, M.; Munro, M. A. R. (1988). Irish tree rings, Santorini and volcanic dust veils. *Nature* **332**/6162. 344–346.

Barberi, F.; Innocenti, F.; Marinelli, G.; Mazzouli, R. (1974). *Vulcanismo c tettonica a placche: esempi nell'area mediterranca. Memiores Societé Geologique* **IV**, **13–2**. 327–358.

Bard, E.; Hamelin, B.; Fairbanks, R.; Zindler, A. (1990). Calibration of the ^{14}C timescale over the past 30 000 years using mass spectrometric U–Th ages from Barbados Corals. *Nature* **345**. 405–419.

Baumann, H. (1982). *Die griechische Pflanzenwelt in Mythos, Kunst und Literatur.* Munich, Germany (Hirmer).

Becker, B.; Kromert, B.; Trimborn, P. A. (1991). Stable-isotope tree-ring timescale of the Late Glacial/Holocene boundary. *Nature* **353**. 647–649.

Betancourt, P. P. (1987). Dating the Aegean Late Bronze Age with radiocarbon. *Archaeometry* **29**. 45–59.

Bietak, M. (1992). Minoan wall-paintings unearthed at ancient Avaris. *Egyptian Archaeology* **2**. 26–28.

Bietak, M. (1996). *Avaris: The Capital of the Hyxos. Recent Excavations of Tell el-Dab'a.* British Museum Press (1996) 98 p. + 34 pl.

Bietak, M. (1997). *The Mode of Representation in Egyptian Art in Comparison to Aegean Bronze Age Art.* International Symposium. The Wall Paintings of Thera, vol. 1. 6–32, London, (The Thera Foundation).

Bietak, M.; Marinatos, N. (1993). The Minoan wall paintings from Avaris. In: Bietak, M. (Ed.) *Ägypten und Levante V.* 49–62.

Bond, A.; Sparks, R. S. J. (1976). The Minoan eruption of Santorini, Greece. In: *Journal of the Geological Society of London* **132**. 1–16.

Buondelmonti. (1824). *Liber Insularum.* Leipzig und Berlin (Sinner).

Boström, K.; Ingri, J.; Boström, B.; Andersson, P.; Löfvendahl, R. (1990). Metallogenesis at Santorini – a Subduction-Zone Related Process. II: Geochemistry and Origin of Hydrothermal Solutions on Nea Kameni, Santorini, Greece. In: Hardy, D. A. (Ed.) *Thera and the Aegean World III.* vol. 2. London (The Thera Foundation). 291–299.

Bott, M. H. P.; Saxov, S.; Talwani, M.; Thiede, J. (eds.) (1983). Structure and development of the Greenland–Scotland Ridge. New methods and concepts. *Nato Conference Series,* Ser. IV: *Marine Sciences.* New York and London (Plenum Press).

Cassiodorus, M. A. *Chronicon.*

Cedrenus, G. (1566). *Annales.* Basel.

Choiseul-Gouffier, C. (1782). *Voyage pittoresque de la Grèce.*

Cita, M. B.; Erba, E.; Lucchi, R.; Pott, M.; van der Meer, R.; Nieto, L. (1996). Stratigraphy and sedimentation in the Mediterranean Ridge diapiric belt. *Marine Geology* **132**(1/4). 131–150.

Cita, M. B.; Camerlenghi, A.; Rimoldi, B. (1996). Deep-sea tsunami deposits in the eastern Mediterranean: new evidence and depositional models. *Sedimentary Geology* **104**(1–4). 155–173.

Clausen, H. B.; Hammer, C. U.; Hvidberg, C. S.; Dahl-Jensen, D.; Steffensen, J. P.; Kipfstuhl, J.; Legrand, M. (1997). A comparison of the volcanic records over the past 4000 years from the Greenland Ice Core Project and DYE 3 Greenland ice cores. *Journal of Geophysical Research* **102** (C12) 26,707–26,723.

Dahl-Jensen, D. *et al.* (1997). A search in north Greenland for a new ice-core drill site. *Journal of Glaciology* **43** No. 144 300–306.

Dansgaard, W.; Hammer, C. U. (1981). Vulkanisme på den nordlige halvkugle registreret i Indlandsisen. *Naturens Verden* **11**. 1–14.

Dapper, O. (1730). *Description exacte des isles de L'Archipel.* Amsterdam.

Davis, E. (1998). Magmatic evolution of the Pleistocene Akrotiri volcanoes. In: *The European Laboratory Volcanoes.* Casale, R., Fytikas, M., Sigvaldason, G. and Vougioukalakis, G. (Eds). Luxembourg, European Commission. 49–67.

Davis, E. N. (1990). A storm in Egypt during the reign of Ahmoe. In: Hardy, D. A. (Ed.) *Thera and the Aegean World III.* vol. 3. London (The Thera Foundation). 232–235.

Davis, E. N. (1997). *The Vapheio Cups and Aegean Gold and Silver Ware.* New York & London (Garland Publishing).

Davis, E. N.; Bastas, C. (1978). Petrology and geochemistry of the metamorphic system of Santorini. In: Doumas, C. (Ed.) *Thera and the Aegean World I.* London (The Thera Foundation). 61–79.

Davies, N, de G. (1930). *The Tombs of Ken-Amun at Thebes.* Publication of the Metropolitan Museum of Art. *Egyptian Expedition.* Vols 1 and 2.

De Capitani, L.; Cita, M. B. (1996). The 'marker-bed' of the Mediterranean Ridge diapiric belt: geochemical characteristics. *Marine Geology* **132**(1/4). 215–225.

Delibasis, N.; Chailas, S.; Lagios, E.; Drakopoulos, J. (1990). Surveillance of Thera Volcano, Greece: microseismicity monitoring. In: Hardy, D. A. (Ed.) *Thera and the Aegean World III.* vol. 2. London (The Thera Foundation). 199–206.

Diapoulis,C. (1980). Prehistoric plants of the islands of the Aegean Sea, Sea Daffodils *(Pancratium Maritimum).* In: Doumas, C. (Ed.) *Thera and the Aegean World II.* vol. 2. London (The Thera Foundation). 129–140.

Dobe, F. (1936). Literaturverzeichnis. In: Reck, H. (Ed.) *Santorin – der Werdegang eines Inselvulkans und sein Ausbruch 1925–1928.* Berlin (D. Reimer). XIV–XXVIII.

Doumas, C. (1974). The Minoan Eruption of the Santorini Volcano. *Antiquity* **XLVIII**. 110–115.

Doumas, C. (1977). *Santorin.* Athens (Delta).

Doumas, C. (1978). Eruptions of the Santorini Volcano from contemporary sources. In: Doumas, C. (Ed.) *Thera and the Aegean World I.* London (The Thera Foundation). 819–823.

Doumas, C. G. (1983). *Thera, Pompeii of the Ancient Aegean.* London (Thames and Hudson).

Doumas, C. G. (1991). *Thera/Santorin. Das Pompeji der alten Ägäis.* Berlin (Koehler & Amelang).

Doumas, C. G. (1992). *The Wall-Paintings of Thera.* Athens (The Thera Foundation).

Doumas, C.; Papazoglou, D. (1977). Santorini tephra from Rhodes. *Nature* **287**. 819–822.

Druitt, T. H. (1985). Vent evolution and lag breccia formation during the Cape Riva eruption of Santorini, Greece. *Journal of Geology* **93**. 439–454.

Druitt, T. H.; Francaviglia, V. (1990). An ancient caldera cliff line at Phira, and its significance for the topography and geology of Pre-Minoan Santorini. In: Hardy, D. A. (Ed.) *Thera and the Aegean World III.* vol. 2. London (The Thera Foundation). 362–369.

Druitt, T. H.; Francaviglia, V. (1992). Caldera formation on Santorini and the physiogeography of the islands in the Late Bronze Age. *Bulletin Volcanologique* **54**. 484–493.

Druitt, T. H.; Mellors, R. A.; Pyle, D. M.; Sparks, R. S. J. (1989). Explosive volcanism on Santorini, Greece. *Geological Magazine* **126**/2. 95–126.

Druitt, T. H., Edwards, L., Lanphere, M., Sparks, R. S. J. and Davis, M. (1998). Volcanic development of Santorini revealed by field, radiometric, chemical, and isotopic studies. In: *The European Laboratory Volcanoes.* Casale, R., Fytikas, M., Sigvaldason, G. and Vougioukalakis, G., (Eds.) Luxembourg, (European Commission).

Durazzo-Morosini, Z. (1936). *Santorin. Die fantastische Insel.* Berlin (Gebr. Mann).

Einfalt, H.-C. (1980a). Chemical and mineralogical investigations of sherds from the Akrotiri excavations. In: Doumas, C. (Ed.) *Thera and the Aegean World II.* London (The Thera Foundation). 459–469.

Einfalt, H.-C. (1980b). Stone materials in ancient Akrotiri – a short compilation. In: Doumas, C. (Ed.) *Thera and the Aegean World II.* London (The Thera Foundation). 523–527.

Eriksen, U. (1990). *Kalkxenolitter fra det minoiske pimpstenslag Santorin, Grækenland.* Aarhus (Geologisk Institut).

Eriksen, U.; Friedrich, W. L.; Buchardt, B.; Tauber, H.; Thomsen, M. S. (1990). The Stronghyle caldera: geological, palaeontological and stable isotope evidence from radiocarbon dated stromatolites from Santorini. In: Hardy, D. A. (Ed.) *Thera and the Aegean World III.* vol. 2. London (The Thera Foundation). 139–150.

Eusebius von Cesararea. (1866). *Chronicorum Canonum quae supersunt.* Berlin (Schoene).

Evans, Sir Arthur (1921–1935). *The Palace of Minos at Knossos.* 4 vols, Macmillan & Co, Ltd, London.

Ferrara, G.; Fytikas, M.; Giuliani, O.; Marinelli, G. (1980). Age of the formation of the Aegean active volcanic arc. In: Doumas, C. (Ed.) *Thera and the Aegean World II*. London (The Thera Foundation). 37–41.

Figuier, L. (1872). *La Terre et les Mers*. Paris.

Flemming, N. C.; Webb, C. O. (1986). Tectonic and eustatic coastal changes during the last 10 000 years derived from archaeological data. *Zeitschrift für Geomorphology* **62**. 1–29.

Fouqué, F. (1869). Une Pompéi Antéhistorique. *Revue des deux mondes* **83**. 923–942.

Fouqué, F. (1879). *Santorin et ses Éruptions*. Paris (Masson & Cie).

Fouqué, F. (1998). *Santorini and its Eruptions*, Translated and commented by Alexander R. McBirney. Baltimore and London (John Hopkins University Press).

Francaviglia, V. (1990). Sea-borne Pumice Deposits of Archaeological Interest on Aegean and Eastern Mediterranean Beaches. In: Hardy, D. A. (Ed.) *Thera and the Aegean World III*. vol. 3. London (The Thera Foundation) 127–134.

Forbes, E. (1844). *Report on the Mollusca and Radiata of the Aegean Sea, and on Their Distribution, Considered as Bearing on Geology*. Report of the thirteenth meeting of the British Association for the Advancement of Science; held at Cork in August 1843. London (John Murray).

Francis, P.; Self, S. (1988). Der Ausbruch des Krakatau. In: *Vulkanismus*. Heidelberg (Spektrum der Wissenschaft) 56–68.

Friedmann, G. M. (1992). Geology illuminates Biblical events. *Geotimes* **3**. 18–20.

Friedrich, W. L. (1966). *Zur Geologie von Brjánslækur (Nordwest-Island) unter besonderer Berücksichtigung der fossilen Flora*. Cologne, Germany (Stolfuss Verlag Bonn).

Friedrich, W. L. (1968). Tertiäre Pflanzen im Basalt von Island. *Meddelelser fra Dansk Geologisk Forening* **18**/3–4 265–276.

Friedrich, W. L. (1980). Fossil plants from Weichselian interstadials, Santorini (Greece) II. In: Doumas, C. (Ed.) *Thera and the Aegean World II*. vol. 2. London (The Thera Foundation) 109–128.

Friedrich, W. L. (1983). Kulturplanter fra Santorins bronzealder. *Naturens Verden* **6–7**. 234–245.

Friedrich, W. L. (1987). Stratigrafi i et vulkansk område – Santorin som eksempel. *Dansk geologisk Forening*. 1–6. Yearbook for 1986.

Friedrich, W. L.; Doumas, C. G. (1990). Was there local access to certain ores/minerals for the Theran people before the Minoan eruption? An addendum. In: Hardy, D. A. (Ed.) *Thera and the Aegean World III*. vol. 1. London (The Thera Foundation) 502–503.

Friedrich, W. L.; Pichler, H. (1976). Radiocarbon dates of Santorini volcanics. *Nature* **262**/5567. 373–374.

Friedrich, W. L.; Símonarson, L. A. (1981). Die fossile Flora Islands: Zeugin der Thule-Landbrücke. *Spektrum der Wissenschaft* **10**. 23–31.

Friedrich, W. L.; Velitzelos, E. (1986). Bemerkungen zur spätquartären Flora von Santorin (Griechenland). *Courier Forschungsinstitut Senckenberg* **86**. 387–395.

Friedrich, W. L.; Pichler, H.; Kussmaul, S. (1977). Quaternary pyroclastics from Santorini/Greece and their significance for the Mediterranean palaeoclimate. *Bulletin of the Geological Society of Denmark* **A26**. 27–39.

Friedrich, W. L.; Pichler, H.; Schiering, W. (1980a). Der Ausbruch des Thera-Vulkans. *Spektrum der Wissenschaft* **9**. 17–24.

Friedrich, W. L.; Friborg, R.; Tauber, H. (1980b). Two radiocarbon dates of the Minoan eruption on Santorini (Greece). In: Doumas, C. G. (Ed.) *Thera and the Aegean World II*. vol. 2. London (The Thera Foundation). 241–243.

Friedrich, W. L.; Eriksen, U.; Tauber, H.; Heinemeier, J.; Rud, N.; Thomsen, M. S.; Buchardt, B. (1988). Existence of a water-filled caldera prior to The Minoan eruption of Santorini, Greece. *Naturwissenschaften* **75**. 567–569.

Friedrich, W. L.; Wagner, P.; Tauber, H. (1990). Radiocarbon dated plant remains from the Akrotiri excavation on Santorini, Greece. In: Hardy, D. A. (Ed.) *Thera and the Aegean World III*. vol. 3. London (The Thera Foundation). 188–196.

Friedrich, W. L.; Seidenkrantz, M. S.; Nielsen, O. B. (in press). Santorini (Greece) before the Minoan eruption – A reconstruction of the ring-island, natural resources and clay deposits from the Akrotiri Excavation. *Geol. Soc. London* in press.

Fritzalas, C. I.; Papadopoulos, G. A. (1988). Volcanic risks and urban planning in the region of Santorini volcano, South Aegean, Greece. In: Marinos & Koukis (Ed.) *Engineering Geology of Ancient Works, Monuments and Historical Sites*. Rotterdam. 1321–1327.

Fytikas, M.; Vougioukalakis, G. (1999). Volcanic structure and evolution of Cimolos-Polyegos. *Bulletin of the Geological Society of Greece* (in press).

Fytikas, M.; Kolios, N.; Vougioukalakis, G. (1990a). Post-Minoan volcanic activity of the Santorini Volcano. Volcanic hazard and risk, forecasting possibilities. In: Hardy, D. A. (Ed.) *Thera and the Aegean World III*. vol. 2. London (The Thera Foundation). 183–198.

Fytikas, M.; Karydakis, G.; Kavouridis, Th.; Kolios, N.; Vougioukalakis, G. (1990b). Geothermal research on Santorini. In: Hardy, D. A. (Ed.) *Thera and the Aegean World III*. vol. 2. London (The Thera Foundation). 241–249.

Fytikas, M.; Vougioukalakis, G. (1993). Volcanic structure and evolution of Kimolos and Polyegos centers (Milos island group). Proceedings of the 6th Congress of the Geological Society of Greece, May 1992. *Bulletin of the Geological Society of Greece* **XXVIII/2**. 221–237.

Galanopoulos, A. G. (1957). The seismic sea wave of July 9, 1956 [In Greek]. *Praktika* **32**. 90–101.

Galanopoulos, A. G. (1958). Zur Bestimmung des Alters der Santorin-Kaldera. *Annales Géologiques des pays Helléniques* **9** 184–185.

Galanopoulos, A. G. (1960a). Tsunamis observed on the coasts of Greece from antiquity to present time. *Annali di Geofisica* **13**/3–4 369–386.

Galanopoulos, A. G. (1981). *New Light on the Legend of Atlantis and the Mycenaean Decadence*. Athens.

Galanopoulos, A. G.; Bacon, E. (1969). *Atlantis. The Truth Behind the Legend*. London (Nelson).

Georgalas, G. C. (1953). L'éruption du volcan de Santorin en 1950. *Bulletin Volcanologique* Sér. II, **XIII** 39–55.

Georgalas, G. C. (1959). L'éruption du volcan de Santorin en 1939–1941. *Bulletin Volcanologique*, Sér. II, **XXI** 3–64.

Georgalas, G. C. (1962). Catalogue of the active volcanoes of the world including solfatara fields. *International Association of Volcanology*, part 12. Rome 1–28.

Georgalas, G. C.; Liatsikas, N. (1926). The eruption of 1925 of Thera's volcano [in Greek]. *Erga A-17*, 15 February 1926.

Gorceix, H.; Mamet, C. (1871). Constructions de l'époque antéhistorique, découvertes à Santorin. *Compte Rendu* **73** 476–478.

Goree, Father. (1712). A relation of a new island, which was raised up from the bottom of the sea, on the 23d of May 1707, in the Bay of Santorini, in the archepelago. *Philosophical Transactions* **XXVII** 354–375.

Greuter, W. (1967). Beiträge zur Flora der Südägäis 8. *Bauhinia* **3** 243–250.

Günther, D. (1972). Vulkanologisch-petrographische Untersuchungen pyroklastischer Folgen auf Santorin (Ägäis/Griechenland). Dissertation. Tübingen. 111 pp.

Günther, D.; Pichler, H. (1973). Die Obere und Untere Bimsstein-Folge auf Santorin. *Neues Jahrbuch für Geologie und Paläontologie* **7** 394–415.

Hallager, E. (1988). Aspects of Aegean long-distance trade in the second millennium B.C. In: *Momenti precoloniale nel Mediterraneo antico. Atti del Convegno Internationale* (Roma 14–16 marzo 1985). *Collezione di Studii Fenici* **28** 91–101.

Hallager, E. (1990). Upper floors in LM I houses. *Bulletin de Correspondance Hellénique*, Suppl. **XIX** 281–291.

Hammer, C. U.; Clausen, H. B. (1990). *The precision of ice-core dating*. In: Hardy, D. A. (Ed.) *Thera and the Aegean World III*. vol. 3. London (The Thera Foundation) 174–178.

Hammer, C.; Clausen, H.; Friedrich, W. L.; Tauber, H. (1987). The Minoan eruption of Santorini in Greece dated to 1645 BC? *Nature* **328**/6130. 517–519.

Hansen, A. (1971). Flora der Inselgruppe Santorin. *Candoella* **26**/1. 109–163.

Heiken, G.; McCoy, F. Jr. (1984). Caldera development during the Minoan eruption, Thira, Cyclades, Greece. *Journal of Geophysical Research* **89**/B10 8441–8462.

Heiken, G.; McCoy, F. (1990). Precursory activity to the Minoan eruption, Thera, Greece. In: Hardy, D. A. (Ed.) *Thera and the Aegean World III*. vol. 2. London (The Thera Foundation) 79–88.

Heiken, G.; McCoy, F.; Sheridan, M. (1990). Palaeotopographic and palaeogeologic reconstruction of Minoan Thera. In: Hardy, D. A. (Ed.) *Thera and the Aegean World III*. vol. 2. London (The Thera Foundation) 370–376.

Hékinian, R. (1988). Vulkane am Meeresgrund. In: Pichler, H. (Ed.) *Vulkanismus* (Spektrum der Wissenschaft) Heidelberg, Germany 92–102.

Helck, W. (1979). Die Beziehungen Ägyptens und Vorderasiens zur Ägäis bis ins 7. Jahrhundert v. Chr. In: *Beiträge der Forschung* vol. 120 Darmstadt (Wissenschaftliche Buchgesellschaft) 1–355.

Heldreich, Th. von. (1899). Die Flora der Insel Thera. In: Hiller von Gaertringen, F. (Ed.) *Die Insel Thera*. vol. 1 (1899) 122–140; vol. 4 (1902) 119–130.

Herodotus (1976). *The History of Herodotus* (transl. from the Greek by Isaac Littlebury). Repr. from the ed. of 1737, London New York (AMS Press) 2 volumes.

Hess, H. (1962). The history of ocean basins. In Engel, A. E. J. *et al.* (Eds.) *Petrologic Studies: A Volume to Honor A. F. Buddington*. Boulder, CO (Geological Society of America) 599–620.

Hiller von Gaertringen, F. (1899). *Die Insel Thera in Altertum und Gegenwart*. Berlin, Reimer Verlag.

Hiller von Gaertringen, F. (Ed.) (1899–1904). *Thera. Untersuchungen, Vermessungen und Ausgrabungen in den Jahren 1895–1902*. Berlin (Verlag G. Reimer).

Hiller von Gaertringen, F. (1936). Besuch der alten Stadt Thera. In: Durazzo-Morosini, Z. (Ed.) *Santorin. Die fantastische Insel*. Berlin (Gebr. Mann) 23–32.

Holmes, A. (1944). *Principles of Physical Geology*. London (Nelson).

Hubberten, H.-W.; Bruns, M.; Calamiotou, M.; Apostolakis, C.; Filippakis, S.; Grimanis, A. (1990). Radiocarbon dates from the Akrotiri excavations. In: Hardy, D. A. (Ed.) *Thera and the Aegean World III*. vol. 3. London (The Thera Foundation) 179–187.

Huijsmans, J. P. P. (1985). Calc-alkaline lavas from the volcanic complex of Santorini, Aegean Sea, Greece. A petrological, geochemical and stratigraphic study (Ph.D. thesis). *Geologica ultraiectina* **41** 316 pp.

Huijsmans, J. P. P.; Barton, M. (1990). New stratigraphic and geochemical data for the Megalo Vouno complex: a dominating volcanic landform in Minoan times. In: Hardy, D. A. (Ed.) *Thera and the Aegean World III*. vol. II. London (The Thera Foundation). 433–441.

Humboldt, A. von. (1845). *Kosmos. Entwurf einer physischen Weltbeschreibung*. vol. 1. Stuttgart and Tübingen (J. G. Cotta'scher Verlag) 252.

Jackson J. (1994). Active tectonics of the Aegean Region. *Annual Review of Earth and Planetary Science* **22**, 239–271.

Johnsen, S. J.; Clausen, H. B.; Dansgaard, W.; Fuhrer, K.; Gundestrup, N.; Hammer, C. U.; Iversen, P.; Jouzel, J.; Stauffer, B.; Steffensen, J. P. (1992). Irregular glacial interstadials recorded in a New Greenland ice core. *Nature* **359** 311–313.

Justinus, M. J. (1972). *Trogi Pompeii Historiarum Philippicarum epitoma*. Stuttgart (Teubner).

Karali-Yannacopoulou, L. (1990). Sea shells, land snails and other marine remains from Akrotiri. In: Hardy, D. A. (Ed.) *Thera and the Aegean World III*. vol. 2. London (The Thera Foundation) 410–415.

Karo, G. (1930). Archäologische Funde. *Archäologischer Anzeiger* **I/II** 135–138.

Keller, J. (1980). Prehistoric pumice tephra on Aegean islands. In: Doumas, C. G. (Ed.) *Thera and the Aegean World II*. London (The Thera Foundation) 49–56.

Keller, J. (1981). Quaternary tephrochronology in the Mediterranean region. In: Self, S. and Sparks, R. S. J. (Eds.) *Tephra Studies* (NATO Advanced Study Institutes Ser. C, vol. 75) Dordrecht/Boston (D. Reidel Publishing Co.) 227–244.

Keller, J. (1982). Mediterranean island arcs. In: Thorpe, R. S. (Ed.) *Andesites*. New York (John Wiley & Sons) 307–325.

Keller, J.; Rehren, Th.; Stadlbauer, E. (1990). Explosive volcanism in the Hellenic Arc: a summary and review. In: Hardy, D. A. (Ed.) *Thera and the Aegean World III*. vol. 2. London (The Thera Foundation) 13–26.

Kircher, A. (1665). *Mundus subterraneus*. Book IV, Chapter 5. Amsterdam.

Klitgaard, K. (1986). Bugten ved Christos. In: Friedrich, W. L.; Andersen, S. B.; Klitgaard, K. (Eds.) *Santorin. Ekskursion og Feltarbejde 1985*. Århus 56–60.

Knidlberger, L. (1975). *Santorin. Insel zwischen Traum und Tag*. München (Hornung Verlag, Viktor Lang).

Knudsen, K. V. (1998). Stratigrafisk undersøgelse af pyroklastiske aflejringer på Santorini, Grækenland. Unpublished MSc thesis, University of Aarhus, Denmark.

Krafft, M.; Krafft, K. (1980). *Volcanoes – Earth Awakening*. Maplewood, NJ (Hammond).

Kraay, C. M. (1966). *Greek Coins*. London (Thames and Hudson).

Ktenas, K. (1926). L'éruption du volcan des Kamenis (Santorin) en 1925, I. *Bulletin Volcanologique* **3** 3–64.

Ktenas, K. (1927). L'éruption du volcan des Kamenis (Santorin) en 1925. *Bulletin Volcanologique* **4** 7–46.

Ktenas, K.; Kokkoros, P. (1928). The parasitic eruption of the Volcano of Kammenis on 23 January 1928 [in Greek]. *Praktika Academy of Athens* **3** 316–322.

Labbeus Bituricus, P. (1670). *Chronologiae historicae*. Paris.

Lacroix, M. A. (1896). Sur la découverte d'un gisement d'empreintes végétales dans les cendres volcaniques anciennes de l'île de Phira (Santorin). *Comptes Rendu Academia Sciences* **123** 656–661.

Lagios, E.; Tzanis, A.; Chailas, S.; Wyss, M. (1990a). Surveillance of Thera Volcano, Greece: monitoring of the geomagnetic field. In: Hardy, D. A. (Ed.) *Thera and the Aegean World III*. vol. 2. London (The Thera Foundation) 207–215.

Lagios, E.; Tzanis, A.; Hipkin, R.; Delibasis, N.; Drakopoulos, J. (1990b). Surveillance of Thera Volcano, Greece: monitoring of the local gravity field. In: Hardy, D. A. (Ed.) *Thera and the Aegean World III*. vol. 2. London (The Thera Foundation) 216–223.

LaMarche, V. C.; Hirschboeck, K. K. (1984). Frost rings in trees as records of major volcanic eruptions. *Nature* **307/1** 121–126.

Lenz, H. O. (1996). *Botanik der alten Griechen und Römer*. 1859. Neudruck Wiesbaden (Sändig).

Leycester, E. M. (1850). Some account of the volcanic group of Santorin or Thera, once called Calliste, or the most beautiful. *Journal of the Royal Geographical Society* **20** 1–38.

Livius, Titus Patavinus. (1977–91). *Römische Geschichte*. München (Heimeran).

Löhr, E. (1971). Verdens ældste træ. *Naturens Verden* 12. 402–410, Rhodos forlag, København.

Luce, J. V. (1969). *The End of Atlantis: New Light on an Old Legend*. London (Book Club Associates).

Lund, M. (1998). Volumenestimering af calderacollapsen under det minoiske udbrud på Santorini, Grækenland. Unpublished thesis, University of Aarhus, 1–101.

Mamet, H. (1874). *De insula Thera*. Insulis (E. Thorin).

Marinatos, S. (1939). The volcanic destruction of Minoan Crete. *Antiquity* **13** 425–439.

Marinatos, S. (1972). *Some Words About the Legend of Atlantis*. Athens (Papachrysanthou).

Marinatos, S. (1974). *Thera VI Colour Plates and Plans*. Athens (Bibliotheke tes en Athenais archaiologikes hetaireios).

Marinatos, S. (1976). *Excavations at Thera VII*. Athens (Bibliotheke tes en Athenais archaiologikes hetaireios).

Marthari, M. (1983). *Excavations at Phtellos, Thera: Period 1980* [in Greek with summary in English]. *Athens Annals of Archaeology* **XV/1** 86–101.

Marthari, M. (1980, 1988). Thera, Phtellos [in Greek]. *Archaeologikon Deltion* **35** (1980), *Chronica* (1988) 472–473.

Marthari, M. (1981, 1989). Thera, Phtellos [in Greek]. *Archaeologikon Deltion* **36** (1981), *Chronica* (1989) 373–375.

Mavor, J. W. (1969). *Voyage to Atlantis.* New York (Putnam).

McCoy, F. W. (1980). The Upper Thera (Minoan) ash in deep-sea sediments: distribution and comparison with other ash layers. In: Doumas, C. G. (Ed.) *Thera and the Aegean World II.* London (The Thera Foundation) 57–78.

McCoy, F. W.; Heiken, G.; (1999). Tsunami generated by the Late Bronze Age eruption of Thera (Santorini), Greece. In Keating, B., Waythomas, C. and Dawson, A. (Eds.) *Special Issue on Tsunami Deposits, Pure and Applied Geophysics,* in press.

McCoy, F. W.; Heiken, G. (2000). The Late Bronze Age explosive eruption of Thera (Santorini), Greece: regional and local effects. In McCoy, F. W. and Heiken, G. (Eds) *Volcanic Hazards and Disasters in Human Antiquity, Special Paper of the Geological Society of America,* in press.

McKenzie, D. (1978). Active tectonics of the Alpine–Himalayan belt: the Aegean Sea and surrounding regions. *The Geophysical Journal of the Royal Astronomical Society* **55/1** 217–254.

McKenzie, D.; Parker, R. L. (1967). The North Pacific: an example of tectonics on a sphere. *Nature* **216** 1276–1280.

Monioudi-Gavala, D. (1997). *Santorini,* Bellonias Foundation.

Moore, J. G. (1966). The 1965 eruption of Taal Volcano. *Science* **151** 955–960.

Morgan, L. (1988). *The Miniature Wall Paintings of Thera. A Study in Aegean Culture and Iconography.* Cambridge (Cambridge University Press).

Morgan, W. J. (1968). Rises, trenches , great faults, and crustal blocks. *Journal of Geophysics Research* **73** 1959–1982.

Murad, E.; Hubberten, H.-W. (1975). Sulfide mineralization in phyllites from the Island of Thera, Santorini Archipelago, Greece. *Neues Jahrbuch für Mineralogie, Monatshefte,* 300–308.

Muratori, L. A. (1746). *Geschichte von Italien.* Leipzig (Verlag Jacob Schuster).

Murawski, H. (1992). *Geologisches Wörterbuch.* 9th edn. Stuttgart (Enke Verlag).

Neumann van Padang, M. (1936). Die Geschichte des Vulkanismus Santorins von ihren Anfängen bis zum zerstörenden Bimssteinausbruch um die Mitte des 2. Jahrtausend vor Christus. In: Reck, H. (Ed.) *Santorin – der Werdegang eines Inselvulkans und sein Ausbruch 1925–1928.* Vol. I. Berlin (D. Reimer) 1–72.

Nicephoros, P. C. (1829 and 1837). *Breviarium rerum post Mauricium gestarum in Corpus Scriptorum histor. Byzantin.* Bonn (Georgios Syncellus).

Niemeier, W.-D. (1990). New archaeological evidence for a 17th century date of the 'Minoan Eruption' from Israel (Tel Kabri, Western Galilee). In: Hardy, D. A. (Ed.) *Thera and the Aegean World III.* vol. 3. London (The Thera Foundation) 120–126.

Ninkovich, D.; Hays, J. D. (1967). Mediterranean island arcs and origin of high potash volcanoes. *Earth and Planetary Science Letters* **16** 331–345.

Ninkovich, D.; Heezen, B. C. (1967). Physical and chemical properties of volcanic glass shards from pozzuolana ash, Thera Island, and from upper and lower ash layers in Eastern Mediterranean deep sea sediments. *Nature* February 11, 582–584.

Oppolzer, T. Ritter von. (1887). *Canon der Finsternisse.* Vienna (Hof- und Staatsdruckerei).

Orosius, P. (1882). *Historiarum adversus paganos.* Vienna (Zangemeister).

Pang, K. D.; Keston, R.; Svrivastava, S. K. (1989). Climatic and hydrologic extremes in early Chinese history: possible causes and dates. *Eos* **70**, 1095.

Papadopoulos, G. A.; Pavlides, S. B. (1992). The large 1956 earthquake in the South Aegean: macroseismic field configuration, faulting, and neotectonics of Amorgos Island. *Earth and Planetary Science Letters* **113** 383–396.

Papastamatiou, J. (1958). Sur l'âge des calcaires cristallins de l'île de Théra (Santorin). *Bulletin of Geological Society of Greece* **3** 104–113 [in Greek with French summary].

Papazachos, B. C.; Panagiotopoulos, D. G., (1993). Normal faults associated with volcanic activity and deep rupture zones in the southern Aegean volcanic arc. *Tectonophysics* **220(1/4)** 301–308.

Pausanias. (1986–89). *Reisen in Griechenland* [*Travelling in Greece*], 3 volumes. Zürich (Artemis).

Pearson, G. W.; Stuiver, M. (1986). High precision calibration of the radiocarbon time scale, 500–2500 BC. *Radiocarbon* **28** 839–862.

Pègues, M. l'Abbé. (1842). *Histoire et Phémènes du Volcan et des iles Volcaniques de Santorin.* Paris (Imprimerie Royale).

Perissoratis, C. (1990). Marine geological research on Santorini: preliminary results. In: Hardy, D. A. (Ed.) *Thera and the Aegean World III.* vol. 2. London (The Thera Foundation) 305–311.

Perissoratis, C. (1995). The Santorini volcanic complex and its relation to the stratigraphy and structure of the Aegean arc, Greece. *Marine Geology* **128(1/2)** 37–58.

Perissoratis, C.; Michailidis, S.; Zacharaki, P.; Angelopoulos, I. (in press). Geologic characteristics of the Santorini caldera and the surrounding area [in Greek with English summary]. *Bulletin of the Geological Society of Greece.*

Petersen, M. D.; Müller, G. (1978). Recent tuffitic sediments around Santorini, Part IV: geochemistry of the iron-rich sediments from the Santorini caldera. In: Doumas, C. G. (Ed.) *Thera and the Aegean World II*. London (The Thera Foundation) 311–322.

Philippson, A. (1896). Die Inselgruppe von Thera. In: Hiller von Gaertringen (Ed.) *Thera. Untersuchungen, Vermessungen und Ausgrabungen in den Jahren 1895–1902*. vol. 1. Berlin (Verlag G. Reimer) 36–82.

Philostratos. (1983). *Das Leben des Apollonios von Tyana* [The Life of Apollonios from Tyana]. Greek and German, edited, translated and commented by Vroni Mumprecht. Munich (Artemis Verlag).

Pichler, H. (1963). Ignimbrite auf Santorin (Ägäische Inseln). *Annales Géologiques des Pays Helléniques* **14** 408–435.

Pichler, H. (1973). 'Base surge'-Ablagerungen auf Santorin. *Naturwissenschaften* **60**/4 198.

Pichler, H. (1988). Einführung (Introduction). In: Pichler H. (Ed.) Vulkanismus: *Natufgewalt, Klimafaktor und kosmische Formkraft*. Heidelberg (Spektrum der Wissenschaft) 7–15.

Pichler, H.; Friedrich, W. L. (1980). Mechanism of the Minoan eruption of Santorini. In: Doumas, C. (Ed.) *Thera and the Aegean World II*. London (The Thera Foundation) 15–30.

Pichler, H.; Kussmaul, S. (1972). The calc-alkaline volcanic rocks of the Santorini group (Aegean Sea, Greece). *Neues Jahrbuch für Mineralogie, Monatshefte* **116**/3 268–307.

Pichler, H.; Kussmaul, S. (1980a). Comments on geological map of the Santorini Islands. In: Doumas, C. G. (Ed.) *Thera and the Aegean World II*. London (The Thera Foundation) 413–427.

Pichler, H.; Kussmaul, S. (1980b). Geological map of the Santorini Islands, (1:20 000). Appendix to: *Thera and the Aegean World II*. London (The Thera Foundation).

Pichler, H.; Schiering, W. (1977). The Thera eruption and Late Minoan-IB destructions on Crete. *Nature* **267**/5614 819–822.

Pichler, H.; Günther, D.; Kussmaul, S. (1972). Inselbildung und Magmen-Genese im Santorin-Archipel. *Die Naturwissenschaften* **59** 188–197.

Pindar. (1958). *Die Dichtungen und Fragmente* [*Poems and Fragments*] translated into German and commented by L. Wolde. Wiesbaden (Limes Verlag).

Platon, N. (1971). *Zakros, The Discovery of a Lost Palace of Ancient Crete*. New York (Charles Scribner's Sons).

Plinius Secundus, Gajus. *Natural History* by Pliny Reprint. London and Cambridge, Mass, 1963–1971, 10 vols. (The Loeb classical library 419) (Original title: Naturalis historia Latin and English).

Plutarch. (1936). *De Pythiae oraculis*. Loeb (The Loeb Classical Library).

Proclus. (1966). *Commentaire sur le Timée, IIe partie: L'Atlantide*. Paris (J. Vrin) Tome premier – livre 1, p. 111–115. Traduction et notes par A.J. Festugière.

Puchelt, H.; Schock, H. H.; Schroll, E.; Hanert, H. (1973). Rezente marine Eisenerze auf Santorini, Griechenland. *Geologische Rundschau* **62**/3 786–812.

Pyle, D. M. (1990). New estimates for the volume of the Minoan eruption. In: Hardy, D. A. (Ed.) *Thera and the Aegean World III*. vol. 2. London (The Thera Foundation) 113–121.

Pyle, D. M.; Ivanovich, M.; Sparks, R. S. J. (1988). Magma–cumulate mixing identified by U–Th disequilibrium dating. *Nature* **331** 157–159.

Pyökäri, M.; Yli-Kyyny, K. (1995). Volcanic beach sediments, their transport, and the development of shore platforms at the base of the caldera wall on the Santorini Islands, southern Greece. *Journal of Sedimentary Research, Section A: Sedimentary, Petrology and Processes* **A65**(2) 436–443.

Quenstedt, W. (1936). Tertiäre und quartäre Mollusken von Santorin. In: Reck, H. (Ed.) *Santorin – der Werdegang eines Inselvulkans und sein Ausbruch 1925–1928*. Berlin (D. Reimer) 73–76.

Raus, T. (1991). Die Flora (Farne und Blütenpflanzen) des Santorin-Archipels. In: Schmalfuss, H. (Ed.) *Santorin: Leben auf Schutt und Asche; ein naturkundlicher Reiseführer*. Weikersheim (Margraf) 109–124.

Reck, H. (Ed.) (1936). *Santorin – Der Werdegang eines Inselvulkans und sein Ausbruch 1925–1928*. Berlin (D. Reimer). 3 vols.

Reiss, W.; Stübel, A. (1868). *Geschichte und Beschreibung der vulkanischen Ausbrüche bei Santorin von der ältesten Zeit bis auf die Gegenwart*. Heidelberg (Bassermann).

Renaudin, L. (1922). Vases préhellèniques de Théra. *Bulletin de Correspondance Hellènique* **46** 113–159.

Renfrew, C. (1985). *The Archaeology of Cult. The Sanctuary at Phylakopi*. London (Thames and Hudson).

Renfrew, C. (1996). Kings, tree rings and the Old World. *Nature*, **381** 733–734.

Richard, F. (1657). *Relation de ce qui c'est passé de plus remarquable a Sant- Erini, Isle de L'Archipel*, Paris.

Ross, L. (1840). *Reisen auf den griechischen Inseln des ägäischen Meeres*. Stuttgart und Tübingen (J.G. Cotta) 1840. Neudruck Halle (Max Niemeyer) 1912.

Sarpaki, A. (1990). 'Small fields or big fields?' That is the question. In: Hardy, D. A. (Ed.) *Thera and the Aegean World III*. vol. 2. London (The Thera Foundation) 422–432.

Sauvage, J.; Jarrige, J.-J. (1978). Sur l'âge des stades initiaux de l'activité volcanique dans l'île de Thira (Grèce): Études palynologiques. *Comptes Rendu Academia Sciences Paris, Sér. D* **286** 929–931.

Schmalfuss, H. (1991). *Santorin: Leben auf Schutt und Asche; ein naturkundlicher Reiseführer*. Weikersheim (Margraf).

Schou Jensen, E. and Håkansson, E. (1991). *Santorini – en aktiv vulkan i Middelhavet*, Varv No. 2, Copehagen.
Schröder, B. (1986). Das postorogene Känozoikum in Griechenland/Ägäis. In: Jacobshagen, V. (Ed.) *Geologie von Griechenland*. Berlin/Stuttgart (Gebrüder Borntraeger) 208–240.
Schultze-Westrum, T. (1963). Die Wildziegen der ägäischen Inseln. *Säugetierkundliche Mitteilungen* **11** 145–182.
Schuster, J. (1936). Pflanzenführende Tuffe auf Santorin. In: Reck, H. (Ed.) *Santorin – der Werdegang eines Inselvulkans und sein Ausbruch 1925–1928*. Berlin (D. Reimer) 77–80.
Schwarzbach, M. (1958). Einige griechische Beispiele zum Kapitel: Bauweise und Erdbebenschäden. *Neues Jahrbuch für Geologie und Paläontologi, Abhandlungen* **106/1** 45–51.
Schwarzbach, M. (1963). Zur Verbreitung der Strukturböden und Wüsten in Island. *Eiszeitalter und Gegenwart* **14** 85–95.
Seidenkrantz, M. S. (1989). Foraminiferfauna fra Akrotirihalvøen. *Georapporter* **11** 22–25.
Seidenkrantz, M. S.; Friedrich, W. L. (1992). Santorini, part of the Hellenic Arc: age relationship of its earliest volcanism. In: Seidenkrantz, M. S. Foraminiferal analyses of shelf areas. Stratigraphy, ecology and taxonomy. PhD. Dissertation, University of Aarhus 41–65.
Self, S.; Rampino, M. R. (1981). The 1883 eruption of Krakatau. *Nature* **294**/5843 699–704.
Seneca, Lucius Annaeus. (1984). An edition with commentary of Seneca, *Natural Questions*, book two Harry M. Hine (Ed.) Salem, N. H. (Ayer) v, 488 pp (Monographs in classical studies) [Original title *Naturales quaestiones*].
Seward, D.; Wagner, G. A.; Pichler, H. (1980). Fission track ages of Santorini volcanics (Greece). In: Doumas, C. G. (Ed.) *Thera and the Aegean World II*. London (The Thera Foundation) 101–108.
Sigurdsson, H.; Carey, S.; Devine, J. D. (1990). Assessment of mass, dynamics and environmental effects of the Minoan eruption of Santorini Volcano. In: Hardy, D. A. (Ed.) *Thera and the Aegean World III*. vol. 2. London (The Thera Foundation) 100–112.
Simkin, T.; Siebert, L.; McClelland, L.; Bridge, D.; Newhall, C.; Latter, J. M. (1981). *Volcanoes of the World*. Stroudsburg (Hutchinson Ross).
Skarpelis, N.; Liati, A. (1990). The prevolcanic basement of Thera at Athinios: metamorphism, plutonism and mineralization. In: Hardy, D. A. (Ed.) *Thera and the Aegean World III*. vol. 2. London (The Thera Foundation) 172–182.
Skarpelis, N.; Kyriakopoulos, K.; Villa, I. (1992). Occurrence and $^{40}Ar/^{39}Ar$ dating of a granite in Thera (Santorini, Greece). *Geologische Rundschau* **81**/3 729–735.
Sparks, R. S. J.; Wilson, C. J. N. (1990). The Minoan deposits: a review of their characteristics and interpretation. In: Hardy, D. A. (Ed.) *Thera and the Aegean World III*. vol. 2. London (The Thera Foundation) 89–99.
Stanley, D. J.; Sheng, H. (1986). Volcanic shards from Santorini (Upper Minoan Ash) in the Nile Delta, Egypt. *Nature* **320** 733–735.
Stommel, H.; Stommel, E. (1985). 1816: Das Jahr ohne Sommer. In: Pichler, H. (Ed.) *Vulkanismus*. Heidelberg (Spektrum der Wissenschaft) 128–135.
Strabo. (1949). *Geography*. London (W. Heinemann Ltd.) Vol. VIII, 203. With an English translation by Horace Leonard Jones.
Strabo. (1954). *Geography*. London (W. Heinemann Ltd.) Vol. V, 161. With an English translation by Horace Leonard Jones.
Strabo. (1961). *Geography*. London (W. Heinemann Ltd.) Vol. IV, 63. With an English translation by Horace Leonard Jones.
Strange, J. (1980). *Kaphtor/Keftiu. A New Investigation*. Leiden (E.J. Brill).
Sullivan, D. G. (1988). The discovery of Santorini Minoan tephra in Western Turkey. *Nature* **333** 552–554.
Symons, C. J. (1888). *The Eruption of Krakatoa*. Royal Society Report of the Krakatoa Committee.
Tarling, D. H.; Downey, W. S. (1990). Archaeomagnetic results from Late Minoan destruction levels on Crete and the 'Minoan' tephra on Thera. In: Hardy, D. A. (Ed.) *Thera and the Aegean World III*. vol. 3. London (The Thera Foundation) 146–159.
Tataris, A. A. (1963). The Eocene in the semi-metamorphosed basement of Thera Island. *Bulletin of the Geological Society of Greece* **6** 232–238. [In Greek with an English summary.].
Tauber, H. (1992). 40 år med Kulstof-14 dateringsmetoden. *Nationalmuseets arbejdsmark*, Copenhagen (National Museum) 144–148.
Televandou, C. (1982). The Mavromatis quarry. *Archaeologikon Deltion (Chronica)*, Athens **37** 358–359.
Theophanes. (1655). *Chronographia*. Paris.
Thorarinsson, S. (1944). *Tefrokronologiska Studier på Island*. København (Munksgaard).
Tournefort, Pitton de. (1707). *Relation d'un voyage en Levant*. Amsterdam 1707 and later editions.
Van Bemmelen, R. W. (1971). Contribution to the geonomic discussions on Thera (Part II). In: *Acta of the 1st International Scientific Congress on the Volcano of Thera*. Athens 142–151.
Vanschoonwinkel, J. (1990). Animal representations in Theran and other Aegean arts. In: Doumas, C. (Ed.) *Thera and the Aegean World III*. vol. I. London (The Thera Foundation) 327–347.
Vaughan, S. J. (1990). Petrographic analysis of the Early Cycladic wares from Akrotiri, Thera. In: Hardy, D. A. (Ed.) *Thera and the Aegean World III*. vol. I. London (The Thera Foundation) 470–487.

Verbeek, R. D. M. (1886). *Krakatau*. Batavia (Imprimerie de l'Etat).
Victor, Sextus Aurelius. (1975). *Historiae Abbreviatae*. Paris (Budé).
Vierhapper, F. (1914, 1919). Beiträge zur Kenntnis der Flora Griechenlands. *Verhandlungen der Zoologisch-Botanischen Gesellschaft Wien* **64** (1914) 239–270; **69** (1919) 102–312.
Vinci, A. (1985). Distribution and chemical composition of tephra layers from Eastern Mediterranean abyssal sediments. *Marine Geology* **64** 143–155.
von Fritsch, K. (1871). Geologische Beschreibung des Ringgebirges von Santorin. *Zeitschrift der Deutschen geologischen Gesellschaft* **23** 125–209.
von Hoff, K. E. A. (1824). *Geschichte der durch Überlieferung nachgewiesenen natürlichen Veränderungen der Erdoberfläche*. II. Theil. Gotha (Justus Perthes).
Von Seebach, K. (1867). Der Vulkan von Santorin. In: Virchow, R. and Holzendorf, F. (Eds.) *Sammlung gemeinverständlicher wissenschaftlicher Vorträge*, 2nd series, Part 38, Berlin.
Von Seebach, K. (1868). Ueber den Vulkan von Santorin und die Eruption von 1866. In: *Abhandlungen der Physicalischen Classe der Königlichen Gesellschaft der Wissenschaften zu Göttingen* **13** Gottingen (Die Dietrichsche Buchhandlung) 1–85.
Vougioukalakis, G.; Francalanchi, L.; Serana, A.; Mitropoulos, D. (1994). The 1649–1650 Kolumbo submarine volcano activity, Santorini, Greece. *International Workshop On European Laboratory Volcanoes, Aci Castello (Catania), Italy. Workshop Proceedings* 189–192.
Walker, G. P. L. (1973). Explosive volcanic eruptions – a new classification scheme. *Geologische Rundschau* **62/2** 431–446.
Walter, H.; Lieth, H. (1967). *Klimadiagramm – Weltatlas*. Jena, Germany (Gustav Fischer).
Warren, P. M. (1978). The unfinished red marble jar at Akrotiri, Thera. In: Doumas, C. (Ed.) *Thera and the Aegean World I*. London (The Thera Foundation) 555–568.
Warren, P. M. (1990). Summary of evidence for the absolute chronology of the early part of the Aegean Late Bronze Age derived from historical Egyptian sources. In: Hardy, D. A. (Ed.) *Thera and the Aegean World III*. vol. 3. London (The Thera Foundation) 24–26.
Warren, P. M.; Puchelt, H. (1990). Stratified pumice from Bronze Age Knossos. In: Hardy, D. A. (Ed.) *Thera and the Aegean World III*. vol. 3. London (The Thera Foundation) 71–81.
Washington, H. S. (1926). The Santorini eruption of 1925. *Bulletin of the Geological Society of America* **37** 349–384.
Wegener, A. (1915). *Die Entstehung der Kontinente und Ozeane*. Braunschweig, Germany (Friedr. Vieweg & Sohn).
White, R. S. & Humphreys, C. J. (1994). Famines and cataclysmic volcanism. *Geology Today* **10** 181–185.
Wiedenbein, F. W. (1988). Quärtärgeologie und Biogeographie der Kykladeninsel Milos. Dissertation Univ. Erlangen-Nürnberg. Erlangen, Germany 1–191.
Wiedenbein, F. W. (1991). Biogeographic effects from tsunamis. *Terra Abstracts* **3/1** 180.
Wijmstra, T. A. (1969). Palynology of the first 30 metres of a 120 m deep section in Northern Greece. *Acta Botanika Neerlandika* **18** 511–527.
Willerding, U. (1973). Bronzezeitliche Pflanzenreste aus Iria und Sinoro. In: *Forschungen und Berichte*. vol. VI Tiryns. 221–241.
Williams, H.; McBirney, A. R. (1968). *Geologic and Geophysical Features of Calderas*. Oregon, USA (Center for Volcanology, University of Oregon).
Wilski, P. (1902). Klimatologische Beobachtungen. In: Hiller von Gaertringen, F. (Ed.) *Die Insel Thera*. vol. 4 Berlin (Verlag G. Reimer) 1–103.
Wilski, P. (1934). Thera. In: Kroll, W. und Mittelhaus, K. (Eds.) *Paulys Real-Encyclopädie der Classischen Altertumswissenschaft*. 2nd series (R–Z), 10th half-volume. Stuttgart (J. B. Metzlersche Verlagsbuchhandlung) 2260–2277.
Wilson, J. T. (1965). A new class of faults and their bearing on continental drift. *Nature* **207** 343–347.
Wilson, L. (1980). Energetics of the Minoan eruption: some revisions. In: Doumas, C. G. (Ed.) *Thera and the Aegean World II*. London (The Thera Foundation) 31–35.
Zielinski, G. A. *et al.* (1994). Record of volcanism since 7000 BC from the GiSP2 Greenland ice core and implications for the volcano–climate system. *Science*, **264**, 948–952.

Index of names

Subject index